Dov Pasternak

AGRICULTURAL PROSPERITY IN DRY AFRICA

Dov Pasternak

Agricultural Prosperity in Dry Africa

Senior Editors & Producers: Contento
Editor: Robin Miller
Design: Liliya Lev Ari
Cover design: Benjie Herskowitz

ISBN: 978-965-550-490-3

International sole distributor: Contento
22 Isserles Street, 6701457 Tel Aviv, Israel
www.ContentoNow.com
Netanel@contento-publishing.com

Dov Pasternak

Agricultural Prosperity in Dry Africa

CONTENTONOW

Table of Contents

List of Abbreviations...9

Foreword 13

Introduction...21

Chapter 1: What is hindering agricultural development
in Africa? 25

Chapter 2: Drivers for development.....................................33

Chapter 3: Strategies for agricultural development............75

Chapter 4: Technologies and crops for rainfed agricultural
development 83

Chapter 5: Irrigation is the solution 145

Chapter 6: Realizing the Potential 189

Conclusions 219

Bibliography..221

Appendixes 227

   Appendix 1: List of vegetable varieties for the three growing
seasons of the Sudano Sahel 228

   Appendix 2: List of successful fruit tree varieties
at the Sadore Research Station of ICRISAT....................229

"What is crucial is not that technical ability but it is imagination in all of its applications."

E. O. Wilson

This book is dedicated to Vesa Jaakola who believed in IPALAC

List of Abbreviations

AVRDC – The World Vegetable Center

AMG – African Market Garden

CAZRI – Central Arid Zone Research Institute

CGIAR – Consultative Group for International Agricultural Research Centers

CIRAD – Centre de Coopération Internationale en Recherche Agronomique pour le Development

CIMMYT – International Maize and Wheat Improvement Center

CLUSA – Cooperative League of the USA

CNSF– Centre National de Semences Forestieres

DG – Director General

ECHO- Educational Concerns for Hunger Organization

EIAR – Ethiopia Institute for Agriculture Research

EU – European Union

FAO-Food and Agriculture Organization (of the United Nations)

FOF – Farmers of the Future

FMNR – Farmer Managed Natural Regeneration

GDP – Gross Domestic Product

GMO – Genetically Modified Organisms

IAAB – Institute for Agriculture and Applied Biology

ICRISAT – International Crops Research Institute for the Semi Arid Tropics

ICRAF – World Agroforestry Center

IER – Institut d'Economie Rurale

IFAD – International Fund for Agriculture Development

IITA – International Institute for Tropical Agriculture

ILRI – International Livestock Research Institute

IMOD – Inclusive Market Oriented Development

INERA – Institut de l'Environnement et de Recherches Agricoles (of Burkina Faso)

IPALAC – International Program for Arid Land Crops

IRAD – Institut de Recherche Agricole pour le Développement

ISRA – Institut Sénégalais de Recherchers Agricole

ITT – International Telephone and Telegraph

MASHAV – The Israeli Centre for International Cooperation

MV – Millennium Village

NAR- National Agriculture Research institute

NGO – Non-governmental organization

NPK – Nitrogen, phosphorus, potassium

NRM – Natural Resources Management

REST – Relief Society of Tigray

R&D – Research and Development

R4D – Research for Development

SAT – Semi-arid tropics

SIM – Serving in Mission

TIPA – Technological Innovation for Poverty Alleviation

UNCCD – United Nations Convention for Combating Desertification

UNESCO – United Nations Educational Scientific and Cultural Organization

UNEP – United Nations Environmental Program

USAID – United States Agency for International Development

USDA – United States Department of Agriculture

Foreword

Whenever I tell friends and colleagues about my experience in international agricultural development, particularly in Africa, they say: "Dov, this is fascinating— you must share your experience and thoughts with the rest of the world." But sharing experiences and thoughts can be very difficult to do when you are busy "trying to save the world."

Now, time, such a rare commodity earlier in my life, has become available, allowing me to sit down at my laptop to share my lifelong experience in international agricultural development.

My major objective in writing this book is to guide those who are involved in the agricultural development of dry Africa about new approaches and strategies that will lead to more successful development.

At the center of my concerns are the interests of farmers.

In this book, I will recount many personal examples of how the lack of understanding of the business of farming, the conditions peculiar to dry Africa and the needs and motivations of farmers are hampering development there.

Many times, I resort to strong language to shake up some parties who, in my opinion, are taking the wrong approaches *without even being aware of the negative consequences that result.* In mentioning this, I would like to apologize to anyone who may feel my criticism is directed at them. This is definitely not my intention.

Even though I am a scientist, by no means have I written this book in the style of a scientific publication. I present here a summary of my more than fourty years of experience in agricultural research and development in Israel, Egypt, Morocco, Argentina and sub-Saharan Africa. It is written in a style that can be understood by a layman but still appreciated by a scientist. The book deals specifically with agriculture in the Sudano Sahel, a region delineated by 300–800 mm/y rain isohyets stretching from Senegal to Eritrea, but it is relevant to other regions of dry Africa that are in the process of transitioning from traditional to more sophisticated forms of agriculture through a process that we call "development".

From 1971 until 2001, I was an agricultural researcher at the Institute for Agriculture and Applied Biology (IAAB) of the Ben-Gurion University of the Negev in Beer Sheva, Israel. My specialty was research on saline water irrigation. Over a period of thirty years, I tested thirty-six crop species for salt tolerance acquiring a unique knowledge of a great variety of crops (field crops, forage crops, vegetables, fruit trees and ornamentals). This experience helped me immensely when

I became the head of Crops and Systems Diversification at the International Crops Research Institute for the Semi-Arid Tropics (ICRISAT) Sahelian Center in Niger.

I headed the IAAB from 1975 until 2001. This relatively small institute with a staff of 100 people has had a very significant impact on the development of the Negev Desert in what is referred to by many as the "Miracle of the Desert." *I believe the primary reason for this success was the absolute academic freedom given to scientists leading to and encouraging innovation.*

My international experience began with a mission to Spain in 1976 that ended with the first introduction of drip irrigation to Europe. Since then, I have conducted consultations and projects in more than twenty countries around the globe. After my first mission to Africa in 1994, I fell in love with the continent and its people, which eventually led to my employment at ICRISAT-Niger from 2001 to 2011. ICRISAT is an international research institute that is focusing on the genetic improvement of sorghum, millet, groundnuts, chickpeas and pigeon peas with outstanding success. I was hired by this visionary organization to conduct research leading to crops and systems diversification.

My period of service in Africa has been the most fertile and fascinating of my professional career; fascinating because of the people whom I met and with whom I lived. West Africans are the kindest people I have ever encountered. How they treat each other with utter politeness and mutual respect, how they look after less fortunate family members, and how they treat the elderly of the family is beyond description. For someone who has been brought up in a competitive society and who has lived under the tension of daily life, the interaction with this highly cultured and peaceful society has been a most enjoyable experience. But the reasons for my satisfaction extend beyond the kindness of the African people. Satisfaction has also come from seeking and finding workable solutions to a vast number of problems impeding African development. From my very first day in Africa, I have been amazed by why so few scientists before me had tried to find these solutions. Perhaps after reading this book, you will understand why this is so.

In addition to conducting a diversified R&D program, I have been able to visit hundreds of villages in eight Sudano Sahelian countries (Cameroon, Niger, Nigeria, Benin, Burkina Faso, Ghana, Mali, Senegal and The Gambia) and to interact with thousands of local farmers, with hundreds of NGOs (non-governmental organizations) and with many other players in the agricultural development arena of Africa. Thus, this book is a result of this intensive experience in agricultural development of West African drylands. It is based mostly on what I have learned from personal experience, from the experiences of close friends and from countless hours of reading and reflecting on what is hindering and what is promoting agriculture in this region.

My commitment to dry Africa began with a program called the International Program for Arid Land Crops (IPALAC) that was conceived by my colleagues from IAAB at Ben-Gurion University. The rationale behind IPALAC is as follows:

Current world agriculture is mainly the outcome of massive crop transfers from one region of the world to another. For example, wheat—the most important crop in the world— originated in West Asia and spread from there to Europe, Asia, Oceania and the Americas. Likewise, the potato from the Andes has transformed the agriculture and eating habits of Europe. Some even say that potatoes were the driving force behind the European "Industrial Revolution." Practically all crops (with the exception of pecans, sunflowers, Jerusalem artichokes and some berries) farmed in the US come from elsewhere. Cassava, the major staple crop of tropical Africa, as well as cocoa, rubber trees and cashews that support the tropical African economy, all find their source in Brazil and its neighbors. On the other hand, Brazil, one of the most prolific agricultural producers in the world, grows many non-native crops: sugarcane originally from New Guinea, soybeans from East Asia, coffee from Ethiopia, corn from Mexico and oranges from Southeast Asia.

The IPALAC team at ICRISAT-Niger: Six scientists, twelve technicians, fifty field workers. Most of the new crops and technologies mentioned in this book were created by this team

To a large extent, this process of crop transfer has bypassed the arid and semi-arid regions of the world, probably because these regions have not been the top priority for development by the colonial powers that were the major players in the global crop transfer movement. Thus, IPALAC was created as a catalyst for crop transfer among the dry regions of the world. The project was originally financed by the Government of Finland, UNESCO the Israeli Agency for International Development Cooperation called MASHAV and by the Brach family from New Jersey.

Map of West and Central Africa: Countries with semi-arid tropics regions are: Cape Verde, Senegal, Mauritania, Mali, Ghana, Burkina Faso, Gambia, Togo, Benin, Niger, Nigeria, Cameroon and Chad

Finland has been by far the greatest supporter of IPALAC and its support has extended over a period of 10 years. Without Finnish support, all that is described in this book could never have been achieved. IPALAC's administrator Arnie Schliessel has been a great help in shaping the direction of this program and in identifying the best partners in Africa. Thank you, Arnie, for this assistance.

UNESCO agreed to support IPALAC on the condition that it concentrated all its efforts in Africa. After two years of operation in Africa as IPALAC coordinator, I realized that for the program to succeed, it would be necessary for me to move there. To accomplish this, I accepted a job offer made by Barry Shapiro from ICRISAT to head a program of crop diversification for the Semi-Arid Tropics (SAT) of Africa, basically-IPALAC. IPALAC donors agreed to transfer the responsibility for IPALAC from Ben-Gurion University to ICRISAT.

And this is how my odyssey with African agriculture began...

Introduction

If you enter the words "African Agriculture" in Google, your search will likely reveal more than 1.5 million entries. So much has been written on this subject by so many people and organizations. Indeed, you can find much of the statistics on African agriculture and its effect on the economy in a voluminous number of articles and books. Therefore, I am making a departure from a usual introduction that begins by giving statistical facts in a background description because this has already been done by so many others. If this is your interest, you can find many of these statistics in FAOSTAT. This introduction will focus instead on the major facts that are relevant for anyone who wants to try his hand in agricultural development in this region. These facts are:

1. Unlike the situation in many parts of the world, African agriculture is practically held in the hands of small land-holders, known as "family-based agriculture", the

size of a family-operated land unit ranges from 0.5 to 3 hectares. The population growth rate of about 3% per year is leading to further diminution of smallholder field sizes.

2. About 60% of sub-Saharan Africa is semi-arid with distinct rainy and dry seasons.

3. About 98% of croplands in sub-Saharan Africa (excluding Madagascar and South Africa) are rainfed.

4. Most households practice full or partial subsistence agriculture.

5. Soil erosion, resulting from land clearing for crop cultivation, inadequate farming practices combined with overgrazing, intensive rains and strong winds is resulting in massive land degradation.

6. In dry Africa, huge reserves of untapped water suitable for irrigation remain unutilized.

7. Soils in most of Africa are very poor with low cation exchange capacity, low amounts of organic matter and low levels of phosphorus and nitrogen. Many soils are acidic, containing large amounts of aluminum, which is toxic to plants.

8. Rains in dry Africa are erratic and fall with very great intensity. Partial or full droughts, causing crops to totally fail are very common.

9. The educational level in rural Africa is very low. In many places, more than 50% of the rural population is illiterate.

10. The population is extremely poor—about a third of the population earns less than $2.0 per day.

11. Organizations dealing with African development generally have very few options for crops, varieties and implementation systems because current choice is very narrow. For this reason, the selection of development actions is often limited.

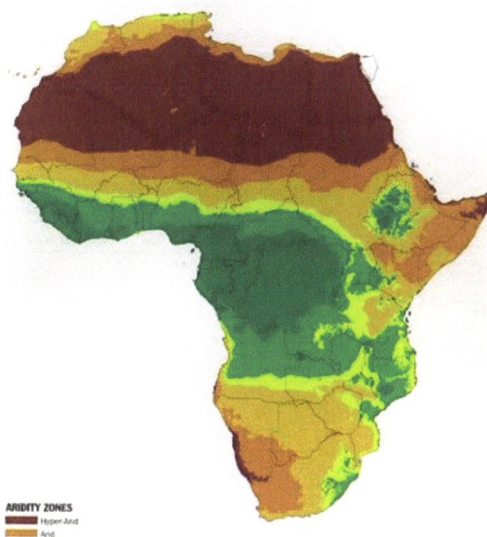

ARIDITY ZONES
Hyper-Arid
Arid

Aridity map of Africa: The wet regions appear in dark green. The other colors represent semi-arid to arid tropics and sub-tropics

To this list, I could add mountains of statistics dealing with so many other aspects complicating and affecting present day life in Africa such as health, sanitation, education, nutrition, social structure and the effects of Colonialism. However, I will not elaborate on these.

It is my hope that after reading this book, you will come away better informed about why agricultural development has been such a challenge in dry Africa and be in a position to help foster the necessary changes which need to occur to allow successful development to proceed.

Chapter 1

What is hindering agricultural development in Africa?

I am starting with the above question because it is in the minds and the mouths of everyone pondering this subject.

In recent decades, developing countries in both Asia and Latin America have made significant strides toward economic prosperity, leaving Africa way behind. People who are unaware of the realities on this continent place the blame for failures directly on the shoulders of Africans, and when discussing agricultural development (by far the major economic activity in sub-Saharan Africa), they place this guilt on African farmers. However, these people could not be more wrong.

The stagnation in agricultural development is not the fault of African farmers.

On the contrary, African farmers are hardworking, honest people, ready (but careful) to accept innovation and eager to improve their lot and that of their family. Rather, this stagnation is caused by enforcing the wrong policies and strategies upon farmers and by a lack of understanding of the natural processes that affect agricultural production in Africa (and elsewhere) by governments and international and regional development agencies as well as NGOs.

I would need to write a separate book to describe all the ways in which many of those responsible for Africa development have and are handicapping it by using wrong approaches and stubbornly adhering to development doctrines, in addition to sheer ignorance. But other authors (see for example Bauer, 1957, Harrison, 1987, Easterly, 2006) have already covered these points in much detail.

Instead, I will give three examples that highlight some of the points that I wish to make. These are examples of projects with great potential for development that have gone astray due to mismanagement by the people responsible for their execution and because of their short duration. I will not discuss many other projects that from their very inception have led nowhere:

1. FAO (UN Food and Agriculture Organization) Multidisciplinary Center for Training and Demonstration of Rural Development

This was one of the first agricultural development projects initiated by the new state of Burkina Faso in the early

1960s. It has been described in a book by Raphael Agmon (1999). The center was constructed near the city of Bobo Dioulassou and contained demonstration plots, training facilities, an agriculture school and a school for training agriculture extension workers. Funding was granted for five years with an option for five additional years if the program were to succeed.

And the program was indeed a success. But during the first five-year period, the government changed, and the new military regime had no interest in and no understanding of agricultural development. Ultimately, the project was discontinued and this huge investment went down the drain, along with the hopes of thousands of farmers.

2. Center for Training on Modern Irrigation Technologies in Burkina Faso

Unlike the project above, funding for this center had no time limitation. It was financed by a group of donors from Texas and was geared towards training young Burkinabe entrepreneurs in intensive horticultural production using drip irrigation.

Its construction took two years. Just as the center was ready to begin operations, it was discovered that the local person responsible for it had mismanaged the funds. This resulted in the withdrawal of the donors from the project. I happened to be there throughout this time period, and I was able to witness the construction of the center and the pullout of the project donors, who had realized that their money and good will had been misused. The center is now deserted. Whenever I drive past it, I have to turn my head away with great pain in my heart.

Durability is yet another factor that determines the success and sustainability of projects. In most cases, donors pre-determine the length of financing for each particular project. Many times, projects are extremely successful while they last but once funding ends, everything collapses. Time and again when this happens, it is the farmers who are blamed. But the blame should actually be placed on the donors and planners of the project, who failed to allow enough time for the farmers to digest the new technologies and approaches and to internally organize themselves.

Each year in the month of September, ICRISAT-Niger traditionally has held two successive "Open Days" to demonstrate its new varieties and technologies. The first day is set aside for "VIPs," meaning government officials and leaders of NGOs. The second day is for farmers and farmers' organizations. On one of these open days, I went into the field to follow up on the demonstration to VIPs by Saidou, our chief technician, of new dual purpose (grain and fodder) cowpea varieties. Saidou was talking to an impressive group of well-groomed NGO leaders. He was explaining the merits of each of the cowpea varieties. From the expressions on their faces, I could tell that this was the first time that these people had ever heard about and seen cowpeas (a major staple crop in the Sudano Sahel), and that they had no understanding of, nor interest in varietal differences. I took over from Saidou and gave the group some basic information about this crop. Unfortunately, even though they now had a better understanding, they still displayed no enthusiasm. **In addition to a lack of understanding, these leaders lacked the passion for agriculture which is at the heart of farmers' livelihoods.**

A few minutes later, the VIPs all entered their air-conditioned Land Cruisers and rushed off to their air-conditioned offices to produce new proposals on agricultural development for their donors to support.

In stark contrast, when we brought farmers to the same field of cowpeas the following day, we were not even given a chance to explain the potential of the varieties. Immediately, the farmers saw the potential by themselves and began running all over the field collecting seeds in their pockets and frocks to try in their own fields. In less than 15 minutes, not a single seed was left on the plants.

3. The first attempt to introduce the African Market Garden to Niger.

In 2002, the drip irrigation company called Netafim and ICRISAT received a grant from the World Bank "Development Marketplace" program to install 100 units of the low pressure drip-irrigation system that I call the "African Market Garden" (AMG) *in just one year.*

Most systems disappeared within five years after the end of the program. The main reasons for this disappointing outcome was that there was not **enough time** to train farmers (mostly women) in the operation of the system, to put in place a company that supplies drip systems and spare parts and to organize the farmer groups.

Another problem is that many times programs have a "second phase," but invariably that second phase does not support the same activities as the first. For example, farmers who began learning to produce vegetables during a "first phase" had to switch to rainfed cereal production in the second phase and, for all intents and purposes, forget all that they had learned about vegetables.

Yet, in spite of all the frustration that farmers have had with so many failed programs, they are still ready to start new projects, anticipating new promises. This is amazing to me.

Fortunately, not all development projects have ended with failure. A USAID project in Niger introduced donkey carts, which have enabled impressive improvements in farming and in rural transportation. This information was given to me by Mark Wentling from USAID, an outstanding representative from this organization. Another USAID project began regional exportation of onions from Niger and now, following livestock, onion exporting is the second largest agricultural export activity in this country.

Here is some advice for donors who have become frustrated with supporting agricultural development in Africa. Please do not despair. You can make a dent in this development. One way to do this is by choosing *experienced* people with farming backgrounds such as Mark Wentling to *lead* development rather than Sociology and Anthropology university graduates and also, hopefully, by following the guidelines given in this book.

William Easterly (2006), in his controversial book, *The White Man's Burden*, reached some of my central conclusions on how aid is damaging development in the Third World. He has divided the players in development into two groups: the "planners" and the "searchers." He further claims that *no planning* is the best strategy. But in his book, he does not give solutions on how to counteract the "planners'" misleading actions. It is a fact of life that the planners *will always be there,* and one must therefore learn to live with them, remembering that most of time the efforts of planners are well-intentioned.

I definitely categorize myself with Easterly's "searchers" and have found out that it is possible to have a significant impact on the development of agriculture in Africa by "flying under the radar." This means that it is sometimes strategic to avoid detection by the planners who are usually the ones in charge of the resources.

Chapter 2

Drivers for development

Before beginning to count the many drivers needed to achieve development, we need to ask ourselves: "Do we really need development?" and if the answer is yes, it should be followed by the question: "What kind of development?"

It is my strong conviction that Africa, particularly semi-arid Africa, needs to jumpstart development because it is lagging so far behind other regions of the world. If improvements in development do not come soon, we can expect hunger, disease, religious fundamentalism, conflicts and mass migration to increase exponentially. A dear friend of mine, Father Godfrey Nzamujo, (who frequently uses the term "flying under the radar"), is the founder of the Songhai Center in Benin. He claims that Africa can no longer afford to undergo a step-by-step economic evolution and believes that development there needs to be accelerated by *leap-frogging*.

Father Godfrey Nzamujo (right) the creator of the Songhai
Center. An outstanding development personality and one of my
mentors. Lennart Woltering (left) is a water resources specialist
from ICRISAT

Agriculture is the major economic activity in dry West dry
Africa. About 80% of the population draws its income from
this source. We are fortunate that there is now international
consensus that economic development in Africa should
start with agriculture. This was not so for many years when
major donors shied away from agricultural development as
though it was a plague. After the FAO World Food Summit
in 1996, it seemed that everyone became intoxicated with
the term "food security," meaning agriculture. Thank God
for small favors.

Unfortunately, the concept of "food security" has boomeranged. Currently, most projects in and for Africa are designed to *increase food production,* guaranteeing food security *by all means regardless of costs and economic considerations.* I deeply dislike this approach for the following reasons:

- The vision of food security means that each person should have access to enough food to maintain his/her vital functions. The problem is that a farmer who reaches the level of "food security" still remains a poor farmer. He still does not have access to all the goods and services with which recent world development has endowed us. Why should we impose such a limited vision on African farmers? What right do we have to tell them to stay poor and be content only with food security? Indeed, they have the same rights as farmers in developed countries. In many of my presentations, in fact, I am suggesting to replace the vision of "food security" with a vision of "Toyota security"—a Toyota 4x4 for each African farmer.

- People do not understand that food in our days and for many years to come is a matter of "accessibility"—not of "availability." Food is available but it is not accessible to the rural poor. During the 2009 drought in East Niger, there was plenty of *imported* food in the markets, but farmers *did not have* the means to purchase that food. Actually, because of the subsistence system, farmers *never* have enough resources to purchase food. If they cannot produce it themselves for whatever reason, they starve. This same situation repeated itself after the 2011 drought.

- I have traveled extensively in dry West Africa, and with few exceptions, I did not see people "dying from hunger." But I have come across many cases where people died from a range of diseases. In Niger, two out of five children born do not reach the age of five because of *disease—not hunger—* and life expectancy in the Sudano Sahel is among the lowest in the world. *Poor health is the number one enemy of Africa—not lack of food* even though the two are closely connected.

Indeed, in dry Africa, the main problem is poverty. If people can significantly raise their income, they can **purchase food**, medicines and medical care, get access to clean water, educate their children and even buy a Toyota 4x4.

For agriculture to prosper, one must treat it for what it is: *a business*, just like any other business in the world. And as a business, its objective is to **generate as much income as possible.**

If farmers can generate a higher income from growing grain, then they should grow grain. But if bananas, tomatoes, coffee, onions, or flowers bring a much higher income than grain, then they should have the right to grow bananas and flowers, sell them and with the money they make, buy grain.

As mentioned above, "food is a matter of accessibility—not of availability." In other words, the assumption that individuals or societies need to produce all their food to be "food secure" is a faulty one. People can buy food from others who are able to produce it more efficiently, provided they have the resources to do so.

Many countries already do this. For example, Israel imports 98% of its grain. Instead of growing grain, its farmers grow horticultural crops, exporting them to Europe. The $1.5 billion earnings per year from these exports, together with other exports, are enough to pay for imported grain. Africa, however, cannot yet afford to import grain as it does not have the resources to purchase it. However, it definitely can focus on products for which it has a relative advantage.

The obvious conclusion from the above deliberation is simple and straightforward.

Let us help farmers to significantly *raise their income* and nearly everything else will fall in place. *This should be the main task of agricultural development in Africa.*

This conclusion has directed my R&D work at ICRISAT towards crop varieties and technologies that maximize income and *create wealth*. However, this is obviously easier said than done.

There are also objective reasons why semi-arid Africa is so poor. Its soils are very infertile and ambient temperatures are quite high. It has long and hot dry seasons and frequent droughts. In addition to these physical constraints, there are numerous man-made constraints that will be further discussed.

In the face of these severe natural and man-made limitations, our challenge is to create an agriculture that will bring wealth and prosperity to its population *against all odds*. This chapter will describe the main drivers for development as well as the major social and political constraints, while suggesting the means to overcome them.

First, we will examine how to change the mindset of people, as this is fundamental in helping to understand the reasons for past failures and the need for change. In Africa (and probably elsewhere), present day thinking—not lack of natural resources— is indeed the *major socio-economic, cultural and political barrier to development*.

1. Changing the mindset of farmers

The need to change the thinking of farmers in Africa is exemplified in the following fascinating story related to me by a friend who worked for an NGO called SIM (Serving in Mission) in the region of Maradi–Niger:

One day I decided to teach a certain village to use zaï holes (planting pits filled with organic matter into which seed is planted) in their sorghum fields. All of the farmers enthusiastically accepted this technique, and they harvested a bumper crop never seen before. This gave me great satisfaction.

I visited this same village during the following rainy season to see how the farmers were progressing. To my surprise, I found no zaï holes in the fields. Moreover, the farmers had not even planted any sorghum using their traditional methods. I approached a group of farmers sitting under the shade of a tree and asked them to explain why they hadn't planted sorghum that year. With broad smiles, they answered proudly, "Well sir, last year, thanks to the zaï technique that you introduced, we produced so much grain that our granaries are still full. So there was no reason to plant sorghum again this year."

In much of Africa, farmers produce to eat rather than sell (subsistence farming). You will never hear such an answer from a market - oriented farmer. His surplus grain would have been sold long ago.

Farmers of the Future learning how to best invest the money they made from selling vegetables

In Senegal, I helped install sixty small units of drip-irrigated market gardens in a three hectare field. During the first dry season, the farmers received a very high yield of onions and revenues were high. I visited the site during the rainy season to find that producers had not planted any crop. Their answer was similar to the answer that the Maradi farmers gave: "Professor, the revenue from the onions was high enough to sustain our families. Why should we bother with another crop during the difficult rainy season?"

The conclusion is straightforward. If you want to alleviate poverty in Africa, you need to change the mindset of farmers from "subsistence" to "market-oriented" agriculture. How could a revolution like this occur within a reasonable period of time was the question that lingered in my mind when I began working on this subcontinent.

In 2003, I was visiting the gardens of schools in northern Mali set up by Mamby Fofana, one of the most creative brains that I know in the development arena of West Africa, when I had an epiphany. In the course of working for a Canadian NGO, Mamby had built botanical gardens in several Primary schools between Mopti and Douendza. He then had given each student a tree to look after and taught them what services they could get from their trees. The children looked after their trees with great care and had started planting "botanical gardens" outside the school compounds. At this place and at this moment, it flashed into my mind that using the primary school system that already exists in most villages of the Sudano-Sahel may be the best way to change the mindsets of farmers simply because the "minds" of children are not yet "set" and are more open to new ideas. I then started the *Farmers of the Future* program at the Sadore village, opposite ICRISAT research center in Niger. Soon after the beginning of this program I realized that childrens' education must be supplemented by the provision of high income options to their parents, particularly the mothers. So I introduced a fruit trees nursery, a BDL field (see page 47), vegetables and fruit trees raising, irrigated cassava, dual purpose (grain/forage) sorghum etc.

Today, twelve years after starting this experiment, the Sadore village has undergone a total transformation. Each woman is now earning about $2,000 per year (as against the average national income of $500 per year). New houses were built and most children go to high school while some continue to university

This is where and when the *Farmers of the Future* (FOF) program was born. It has taken me seven years of trial and error to develop the principles of this program in the Sadore village.

Fortunately, in 2011, a Canadian NGO called "Pencils for Kids," an American NGO called "Eliminate Poverty Now" and Libo a local NGO decided to finance and actively participate in pilot testing the FOF in three villages near Niamey, the capital of Niger.

The FOF program includes a vegetable garden, a fruit tree plantation, and a fruit tree nursery and a shed for small animal fattening, all managed by mothers and jointly operated by mothers and children.. All these are *income generating* activities. A school curriculum was developed, and children receive both theoretical and practical lessons on all these technologies. The children are taught about adding value through storage of cowpeas and other products, short- term fattening of animals and grafting fruit trees. They learn how to produce high value vegetables and fruit trees with irrigation. The central topic of the program is *economics* where mothers and children are being taught about using the income generated for savings, credit and investment.

Fathers are growing rain-fed watermelons and improved cowpea varieties using the zaï system and mothers are practicing vegetable production and fruit tree nurseries.

On paper, this all sounds promising, but we will need to wait a few years to summarize the lessons learned from this pilot program. If it succeeds, we intend to convince governments and funding organizations to adopt this approach and implement it throughout the entire Sahelian region and beyond.

The *Farmers of the Future*, that changes mind set of children and parents, just might be the "leap-frogging" action of Father Godfrey that we are seeking leading to the creation a new generation of farmers.

2. Changing the mindset of NGOs and individuals

A major problem with NGOs and influential individuals is that they are often guided by principles which may be inerrant or conflicting. For example, two opposing doctrines, now in furious competition with each other, concern genetically modified organs (GMOs). NGOs originating primarily in Europe are actively fighting the introduction of genetically modified plants to Africa because of fears that they will contaminate the land and result in environmental disaster. On the other hand, there is avid promotion of GMOs, predominantly by Americans, who view them as the key to solving all the problems in African agriculture (Paarlberg, 2009). At this point in time, there has been no proof that GMOs cause more harm than conventionally bred plants.

Another issue causing controversy is the doctrine of organic agriculture. Organic agriculture bans the use of toxic insecticides and chemical fertilizers. While there is good reason for banning toxic insecticides, as they can be harmful to people if operational instructions are not carefully followed, in Africa, very few "organic" insecticides or fabricated natural enemies are not readily available, and often those that are, tend to be very expensive. Nevertheless our experience with spraying with extracts of neem leaves has been very successful. The neem tree that originated in India is now spreading all over the Sudano Sahel. Thus not only is the neem leaves spray quite effective, it is also very inexpensive.

Organic agriculture also bans the use of fertilizers. Unlike the case for banning toxic chemicals, which I basically accept, I cannot understand the wisdom behind banning fertilizers in Africa. Yes, overuse of fertilizers can contaminate soils and waterways. But so do cow effluents that are cherished by the organic agriculture fans (with the added contamination of the atmosphere by greenhouse gases). But who in Africa is using excess fertilizers? On the contrary, fertilizer use should be more widespread.

Another developmental doctrine is the use of only traditional technologies and traditional plant varieties under the assumption that farmers know best what is good for them and are already employing the best crops and technologies. Unfortunately, this concept is based on ignorance. Yes (and you will see many examples in this book), farmers have developed many beneficial technologies and varieties that are superior to alternative solutions, but science also plays an important role in agricultural development, and it has made a far more substantial contribution than have indigenous farmers technologies.

This latter doctrine is interwoven with the fundamentalist idea to do only what farmers suggest. In its classical model, this means going to the farmer, asking him what he wants and then implementing a development plan based on his wishes. NGOs that follow this approach apparently are not aware that most African farmers do not read books and do not use the Internet. Nor are they otherwise informed of innovation, and that the furthest destination that most have ever visited in their lives is the next village. So how local farmers could be expected to dictate to university graduates and international experts which development strategies should be utilized? This policy was created as a reaction to the top-down planning of production activities by governments and development agencies. However, it is also representative of the fact that the farmer is more knowledgeable about agriculture than many leaders of large NGOs, who have come to Africa to "help" the farmer.

Unfortunately, farmers' choices of options are currently extremely limited. *It is the duty* of the development agency to bring a range of options to farmers, encouraging them to experiment under local farming conditions and then to decide which option seems best. NGOs must remember that they exist to help rural people to overcome poverty. This is their main task, and they should not harm farming progress with idealistic doctrines.

3. Changing the mindset of research organizations

Following the successes of the 1960's in developing high yielding varieties of wheat and rice, the Consultative Group for International Agricultural Research Centers (CGIAR), an umbrella organization of fifteen international research centers was formed. They decided that most of their institutes should concentrate on crop breeding because they believed plant breeding would yield the highest output for investment in research. The history of CGIAR and its contributions are given by Alston, et al (2006).

CGIAR has allotted a limited number of crops that each of their centers are permitted to work with the so-called "mandate" crops. Any institute dealing with a given number of crops is not allowed to independently add other crops to their mandate list or become involved in research on crops mandated to other institutes. I was told that in the heyday of ICRISAT-Niger, scientists that became involved in non-mandated activities were fired. I, myself, was fired because after fighting for ten years against CGIAR doctrines, I became a bit too careless, took too many "risks" and was caught by the new ICRISAT director for West Africa, a diligent follower of the CGIAR doctrines. And poof. I was blown out of the system along with the crops and systems diversification program that had been the central activity of ICRISAT-Niger for ten consecutive years.

When I first visited the ICRISAT Sahelian Center near Niamey, Niger, the best equipped research center in all of the Sahel, I was amazed to find out that this center was involved *only* with millet breeding and soil fertility. There was no research underway to examine and find solutions to central problems of agricultural production in the Sahel, such as halting land degradation (not just "analyzing" it), irrigation, vegetable cultivation (the mandate of another institute) or cultivation of fruit trees or other income-generating field crops such as cowpeas and cassava that are the "mandate crops" of yet another CGIAR institute, the International Institute for Tropical Agriculture (IITA).

While it is true that CGIAR has created additional centers focusing not only on crop breeding but also on irrigation research, animal production, agroforestry, forestry, economics and fisheries covering much of the range of agricultural activities (with the noted exception of horticulture), these initiatives *were missing* at the Sahelian center, where they were needed most.

It has already been demonstrated that plant breeding of selected crops, as important as this is, will not bring solutions to the vast array of agricultural problems in any particular locale, particularly not in dry Africa with so many other limiting factors.

The best solution for effective agricultural research by CGIAR is to create ecology-based multidisciplinary research centers: for instance, a center for the semi-arid tropics, a center for mountain agriculture and a center for tropical agriculture. Each center would deal with a vast array of agricultural disciplines without assignations to mandate crops. For example, cassava does well in tropical regions but it does better in semi-arid regions if irrigated, and potatoes— a central crop in the Sudano Sahel that currently is a mandate crop of yet another CGIAR center situated in far away Peru.

In order to overcome this limitation, I convinced ICRISAT and other relevant institutes to build a multidisciplinary team through joint appointments. The crops and system diversification team that was consequently built at ICRISAT-Niger was composed of joint appointments (formal and informal) with the World Vegetable Center (AVRDC), the World Agroforestry Center (ICRAF), the International Livestock Research Institute (ILRI) and the Centre de Coopération Internationale en Recherche Agronomique pour le Développement (CIRAD). In addition, we cooperated with the International Institute of Tropical Agriculture (IITA) for cowpea and cassava research. To this team, ICRISAT added a "systems agronomist" (me) and a soil scientist. An ICRISAT economist who advised us in all our R&D activities completed this integrated team. This is an example of how to solve a problem "against all odds". The outcome of this collaboration was a wealth of scientific and technological breakthroughs. I am humbly suggesting that the CGIAR should consider re-creating examples of this unique and successful collaboration, leading to the eventual formation of the proposed multidisciplinary research centers.

CGIAR research is ruled by what it is now called the "Science Council" (formerly the Technical Committee). The Science Council is composed of distinguished scholars from international universities and research centers, who meet and decide on the priorities for the CGIAR centers. However, the priorities they establish are the same for all 15 centers. This is a faulty approach since each region of the developing world should have its own priorities.

Whenever the Science Council detects a diversion from the priorities they set, they affix the derogatory term, "mission drift" to describe it. As learned scientists, these council members, should be aware that "mission drifts" as conceived by such luminaries as Galileo, Darwin, Einstein and others are what have led to breakthrough discoveries in the field of science. On a personal level, what upsets me as a Professor Emeritus is the fact that Science Council members are dedicated fighters for academic freedom in the universities from which they come. But once assigned to their council positions, a transformation seemingly occurs, and they suddenly fear academic freedom.

On a positive note, ICRISAT, very recently, has adopted a new strategy that it calls Inclusive Market-Oriented Development (IMOD). This strategy recognizes the fact that agriculture in the semi-arid tropics is an integrated system encompassing not only rain-fed grains and legumes but also livestock, agroforestry, soil fertility, crops and systems diversification, irrigation and value chains. It also states clearly that the alleviation of poverty is the major concern of its Research for Development (R4D) strategy.

This is a good model for the rest of CGIAR to follow, but based on the past record of such research institutes, scientists find it very difficult to step out of their comfort zones. Therefore, I am doubtful that ICRISAT will be able to carry out this new vision in the manner in which it was intended.

At this point, I would like to depart a bit from discussing agriculture only in Africa.

I have also had the opportunity to advise Argentina, Chile, Paraguay and Haiti on development of agriculture in their dry regions. After working with the CGIAR system for ten years, I realized that ICRISAT and other CGIAR centers have amassed a significant knowledge base for agricultural development in the semi-arid tropics of Africa and Asia. The semi-arid tropics of South and Central America (perhaps with the exception of Brazil, which is well advanced in cultivating its semi-arid lands) could and **should** benefit from this know-how, and *it is the responsibility of the CGIAR, its supporting organizations and Latin America countries to find the means to transfer the expertise created in Africa and Asia by the CGIAR centers, particularly ICRISAT to America's semi-arid tropics. Relatively little investment in the transfer of this know-how should result in very high* returns in a short period of time. For example, in a recent mission to Haiti, I found that groundnuts, sorghum and pigeon peas are their three major rainfed crops. However, the yields and quality of the local varieties of these crops is very poor because they are using very old traditional varieties. Transfer of ICRISAT's high yielding and high quality varieties to Haiti should result in significant benefit to this country.

4. Changing the mindset of governments and development agencies

In 1994, I spent a few weeks in northern Cameroon preparing a plan for the Minister of Agriculture for the development of the North and far North regions of this country. One day, my hosts from the Ministry of Agriculture took me to a rice irrigation project that was using water from one of the tributaries of the Logon River. The soil at this site was very heavy and compact, and only powerful tractors could plow the land in preparation for planting rice. At that time, only one tractor had survived out of the armada left by the French who had constructed the dam and the irrigation network.

After analyzing the situation, I proposed that alfalfa production at this site could be more profitable than rice. The region has plenty of livestock and, hence, there is local demand for forage. In addition, alfalfa is a perennial crop, so this would eliminate the need to plow the land every year as with rice. To test this idea, I recommended planting a pilot field of alfalfa in my report.

After reading the draft report, the local hosts came to me very frightened: "Please Professor Dov," (this is what they call me in West Africa) they said, "do not mention alfalfa in your report. Mention only rice and perhaps other irrigated cereals or pulses. If we allow you to recommend alfalfa, the Minister will fire us."

In the recent past many African leaders have not used the development money properly.

Fortunately the attitude of African governments and politicians towards development has been changing in the last few years. Governments are now setting their demands from development banks and other organizations dealing with development instead of passively accepting their dictates. They are now requesting financial support for aspects of agricultural intensification including irrigation and building of value chains. My suggestion to grow alfalfa in Cameroon—the idea that was strongly rejected in 1994—will now be accepted there with open arms. Still African politicians have a long way to go.

Those involved in international development are very much aware of the "fashions" in agriculture development established by governments and development agencies.

These fashions are basically *doctrines* set up by some consensus of development agencies, and they rule the shape and direction of development until a new fashion takes over. Without conducting a study on the reasons for these fashions and on their dynamics, my educated guess is that development bureaucrats feel secure operating "within the box" of the current trend in order to protect themselves from failure. "We did it because everybody was doing it." Another possible reason for reliance on fashions is the fact that many officials involved in development are not usually fluent in agricultural development, so they look for solutions in the strategies of other development agencies.

Here are some of the fashions/buzzwords that have dominated the development arena for various periods of time:

- *Food self sufficiency*
- *Agriculture is not a priority for development*
- *Combating desertification*
- *Food security*
- *Poverty alleviation*
- *The gender issue*
- *Improved nutrition*
- *Access to safe water*
- *Resilience to climate change*
- *Governance*

The "agriculture is not a priority" doctrine has been the most damaging not only because it was wrong but also because it lasted too many years, resulting in significant stagnation of agriculture and, hence, economic development. While this fashion ruled, it was considered heretical to use the word "agriculture." One could not mention it in proposals to donors. "Agriculture" was cleverly replaced by the politically correct term: "Natural Resources Management" (NRM). I well remember using NRM to mean agriculture in the early 1990s in some of my proposals to USAID.

Whenever a new fashion is adopted, all previous ones lose importance. Now, as I am writing these words, we are in the era of "Resilience to Climate Change." The focus of everyone's work has now shifted to combat the outcomes of this change. Suddenly, food security, poverty alleviation and the multitude of other problems are seemingly *no longer important. It is as though these problems have already been solved.*

I am at a loss to suggest what needs to be done to alter this kind of thinking by development organizations. These are the kinds of odds that one needs to fight to reach a prosperous agriculture in Africa. *However, it is a certainty that changing the mindset of governments and development agencies is much more difficult than changing the mindset of farmers.*

5. Changing the mindset of all on the issue of sources of energy food

Governments, international donors, NGOs and research organizations working in semi-arid Africa truly believe that cereals are the *only* source of energy food. The result is that all of these entities encourage concentration on grain crops such as rainfed cereals and irrigated rice in development projects and in research. This mindset is harmful for development in dry Africa, and my hope is that this particular section will be read carefully and digested by all of my colleagues in development.

In general, it is not well known that *dates have more than 3,000 calories per kilogram compared to cooked rice with 1,800 cal/kg.* In addition, dates are an arid land crop. Because they ate lots of dates, Iraqis did not suffer hunger during the embargo imposed by U.S. President George H. W. Bush.

The potential of bananas under irrigation as an energy crop should also be explored in semi-arid Africa. Although bananas provide "only" 970 cal/kg (half as much as cooked rice), in Uganda about 70% of the calories consumed by its inhabitants are from bananas and plantains. Irrigated bananas can easily give a yield of twenty tons of fruit per hectare per year compared to irrigated rice that under double cropping (rainy plus dry season) and intensive production yields up to only eight tons/ha per year. Moreover, the market price of 1 kg of bananas in dry West Africa is double the market price of rice.

Another staple food source in tropical Africa is the "queen" called cassava that is already supplying much of the calories of the population. Biafrans did not suffer hunger as much as expected while surrounded for months by the Nigerian army because of the availability of cassava. Whereas in the tropics, the crop is rainfed, in the Sudano Sahel, it should be irrigated. We tested twenty-seven selected IITA cassava varieties under irrigation and identified varieties for cooking that yield fourty tons per hectare. One processing variety (used for flour production) even gave 69 tons/ha. The following simple calculation exemplifies the worth of say 30 tons/ha of cassava in dry West Africa compared with the value of 8 tons/ha of double cropped irrigated rice:

One kg of cassava roots is sold in dry West Africa for $1.5 and bought from farmers at $1.0. Therefore, the gross revenue of thirty tons of cassava (the yield per hectare) is a respectable $30,000. In contrast, rice yield is 8 tons/ha, and the price of one kg of rice in dry West Africa is 65¢. Therefore, the gross revenue from one hectare of *irrigated* rice is only $1,300 (note, $1,300 versus $30,000). As the calorific value/kg of the two crops is similar, is there a reason that farmers should miss the opportunity to gain these profits while feeding the rest of the population? Can someone please tell me why cassava is desired as a crop by development agencies in tropical Africa but is taboo in dry Africa?

Whenever governments and development agencies devise new strategies, they often try to impose them on farmers, many times in an insensitive manner. For example, a government agency for irrigation using Niger River water supplies water to farmers who are obliged to grow rice and *only* rice with the water they receive. When some farmers started to grow onions after rainy season, this agency cut off their water supply.

Knowledgeable planners have also decided that today, Senegal needs to grow more rice. They are beginning to impose this strategy on farmers in a number of ways *without consulting if this is what farmers want to do and without asking themselves if this doctrine will bring higher income to farmers than other existing alternatives.*

The farmers of Africa *should stop allowing the planners to make decisions for them.* They need to unite and fight back, *as the planners have no right over their sweat and blood. Farmers should be the ones to determine their own future.*

Furthermore, government and development agency bureaucrats should stop imposing their doctrines on farmers. The era of the Soviet-planned economy ended simply because it did not work, and the Gulag, where Stalin used to send "non-obedient" farmers, is closed. Enforcement actions nearly always create opposite results to those intended—resulting in much lower food production.

6. Adding value

Adding value to a certain agricultural product (if this is done by the farmer) can significantly increase its value and hence a farmer's income. Many think that processing increases the value of a product. But very often this is not true. Selling the product when it is fresh, when markets are accessible and prices are reasonable, brings much higher income then selling the fresh product for processing. There are some exceptions though. For example, in Sahelian Africa the revenue from dried sweet pepper is higher than the revenue from the amount of fresh sweet pepper used to produce the dried sweet pepper.

Product storage is another way to add value. For example Pasternak et al (2013) demonstrated that storing onions in an effective way for a period of six months can result in a ten-fold increase of the onion price.

7. Innovation

Everyone is aware that innovation is a major driver of economic development. The amazing worldwide economic progress over the last two centuries has resulted from innovations that began with the invention of the steam engine and continue today in the proliferation of the Internet and information technology.

In 2006, I identified about 40 innovative farmers through NGOs and the government extension service, with the objective of creating a Niger "Innovators Club." The room in which these farmers assembled was full of electricity. The innovators immediately created friendships among themselves and started telling each other about their new technologies and the difficult paths that they had had to take in order to test, refine and prove the worth of their innovations. I described the many innovations that ICRISAT was working on to them, and they almost unanimously requested our help in trying to test those new technologies.

Unfortunately, due to a lack of funds (please find me a donor who will give money to fund an innovators' club), I had to stop supporting the Innovators Club of Niger. But I am still convinced that this is an excellent tool for creating, testing and disseminating innovations and that bringing innovator farmers together will have a multiplying effect. I continue to practice this approach in my current position as agricultural development adviser.

Everybody talks of and praises innovation but, in reality, very few actually support its *creation* and promotion. There are many reasons for this rejection. One is fear that innovation will collide with existing doctrines. Another one is that innovation will physically harm existing practices that so many are benefiting from. Yet perhaps the most important reason for rejection is that people are conformists by nature; they are afraid of going against convention.

The following is an example of an initiative of a single visionary that caused a breakthrough in the development of the greenhouse industry in Israel. In 1970, a greenhouse tomato grower from Auckland, New Zealand named Eddie Peretz visited the Western Negev desert of Israel and decided that this region was most suitable for the production of greenhouse tomatoes for export to England. Eddie was able to convince Dr. Raanan Weiss, the person in charge of agricultural settlement in Israel, to provide him with a small greenhouse in which, against strong opposition by the "experts," he (and his wife Nancy) demonstrated the feasibility of the production of tomatoes for export in greenhouses.

Due to their success, today, thousands of vegetable greenhouses have sprouted all over Israel. Intensive R&D followed the first "discovery" by Eddie Peretz and resulted in the creation of large aerated plastic houses, low cost cooling systems, special polyethylene sheets, insect nets, formulas for plant nutrition and adapted greenhouse varieties that facilitated the boom in greenhouse vegetable proliferation.

If one wants to change a stagnant situation, he or she must try new ways (i.e., innovation) than those that have led to the existing stagnation. Or as Albert Einstein phrased it: *"You cannot solve a problem using the same approaches that created the problem."*

Innovator-farmers are a venue for the creation and dissemination of new technologies. In nearly every village in Africa, one encounters them. These farmers are a special breed. They are non-conformists who swim against the mainstream by trying new technologies that they have either invented or adopted from other places. These people are appreciated but are not always well-liked by their communities. Many times, their communities actually try to hinder their efforts. But once these innovators prove that a technology works, many others will start copying them.

There is no magic formula for the creation of innovation, and not everyone can be an innovator. But innovation can be encouraged and nurtured. Scientists from American and Israeli universities where *academic freedom* prevails receive far more Nobel Prizes in Science than scientists from other countries. *Freedom of thought and freedom of action are central conditions that must exist to give birth to innovation.*

Many times a person or group of people promote an innovative technology to solve a particular problem, but ultimately, this technology proves to be much more valuable in solving problems *not envisioned by the inventors.*

Here is an example:

The inventors of drip irrigation in Israel tried to produce a system that would save water. Thus, they buried the drip laterals in the ground to eliminate evaporation from soil and plant surfaces.

Beginning in 2002, I tried to introduce an irrigation system that I developed and called the "African Market Garden" (AMG) to West Africa. My main objective was to increase yields and save energy and water. But after I installed the first systems in some Sahelian countries, I realized that in West Africa, **labor saving** (not water saving or higher yields) —is the main advantage of drip irrigation. This is because most vegetable gardens in this region are hand irrigated, and the labor input for irrigation is enormous.

Yehuda Zohar, a Ministry of Agriculture extension worker, became aware that drip irrigation had benefits beyond just water conservation. By supplying water and soluble fertilizer to individual plants on a daily basis, crop yields can be significantly increased in comparison with other irrigation systems, particularly in non-fertile soils. In 1964, Zohar installed 1,000 m² drip-irrigated plots in four agriculture settlements in the dry Arava Valley of Israel. The main idea was still to save water and, therefore, the drip laterals were buried in the ground in three out of the four settlements. In the fourth settlement (Yotvata near the Red Sea), the trencher that was supposed to dig the trenches for burying the drip laterals did not arrive on time, so instead the laterals were spread on top of the ground. Surprisingly, results were better than those of the buried drippers because there was no clogging of drippers by plant roots. This is how drippers came up to the surface and how the original objective of water saving became secondary, at least in the non-fertile soils of the Arava Valley of Israel.

This development happened at the exact time I finished my first University degree in Agriculture and had moved to one of the above four settlements called Ein Yahav to conduct vegetable research under the guidance of scientists from the Vegetable Department of the Volcani Center—the Israeli National Agriculture Research Institute. Yehuda Zohar had asked the farmers of Ein Yahav to look after the drip-irrigation plot, but they refused because they were busy "growing vegetables with proven existing methods." Yehuda then turned to me, and I agreed to look after his drip-irrigated field.

Volcani scientists instructed me to conduct a series of trials on vegetables using sprinkler irrigation. The sprinkler-irrigated plot was adjacent to the drip-irrigated plot. No matter what I did, the sprinkler-irrigated plants *just refused to grow*. In contrast, the drip-irrigated plants grew like Jack's magic beans. When the tomatoes from this plot were ready for harvesting, I asked the farmers of Ein Yahav to come and help because I was unable to master the fruit load on these plants by myself. The farmers worked steadily for one month to harvest the tomatoes and sweet pepper fruit grown by drip irrigation. It was then and there that Arava Valley farmers decided to adopt this technique, and the rest is history.

But drip irrigation had yet another *unexpected* advantage. The water of the Arava Valley is saline. The results in the Ein Yahav fields that were irrigated with saline water showed that drip irrigation can be an effective technology for irrigation with saline water. Six years later, after returning to Israel from Australia with an M.Sc and a PhD in agriculture, I dedicated the next thirty years of my life to study how to use saline water for irrigation by means of a drip system.

Some innovations do not require outside investment to promote their proliferation.

The zaï hole, already mentioned, was invented by farmers in Burkina Faso, and this technique started spreading in this country with very little external help. One day an NGO from Niger brought a group of farmers from the Tahoua region in north Niger to Burkina Faso to inspect the zaï hole technology. Upon returning to Niger, Tahoua farmers started digging zaï holes, and it also became a common technology in this region (Hassan 2006). One trip to Burkina Faso was all that was needed for zaï to be adopted in Niger.

Professional Zai holes in Illela-Niger

Another example of an easy success is the case of introduction of new cowpea varieties to two marginal regions of Niger. In 2008, we at ICRISAT introduced the ISV 20 and ISV 128 cowpea varieties to northern Tahoua and to the Oualam region. These are dual purpose high-yielding, drought and pest tolerant varieties giving high grain and high forage yields. Within three years, without any external help, these varieties pushed aside all other varieties that were growing in these regions.

A good example of grass roots adaptation of a new crop is the case of the introduction of dolich (a crawling legume) to the Tahoua region of Niger by former Niger President Mamadou Tanja (while he was the governor there) in 1984. Tanja simply distributed dolich seeds to a few farmers. Subsequently, this crop spread throughout this region just from farmer-to-farmer contacts. Today, it is probably more important to Tahoua as a source of protein and forage than cowpeas, its botanic relative that grows in the non-secure rainy season.

Another innovative idea known as the Millennium Village (MV) has been conceived by Prof. Jeffrey Sachs of the Columbia University in New York. The basic concept of the MV is that development should be carried out in an integrated manner meaning production, value chain development, education, health and processing occurs all at once. This approach was used in Israel in the 1950s and 1960s when massive settlement of the country took place (Weitz and Rockach, 1968). However, the MV requires heavy investment and time to prove its feasibility. Many are, therefore, asking if this is indeed the right approach.

In answer to this question, I tell this joke:

A couple came to a rabbi asking him to arrange a divorce for them. First, the rabbi asked the woman to enter his room. "My husband," she said, "is a terrible man. He is dirty, he drinks, he does not bring money home" and so on.

"You are right," the rabbi told her, and she left the room.

Then the husband entered and complained about his wife: "She does not cook, she neglects the children, she uses all my money on herself," and so on.

The rabbi told the man, "You know what? You are right."

After the man left the room, the rabbi's wife who was present the whole time asked him, "My dear how could both the wife and the husband each be right?"

"You know what my dear wife?" answered the rabbi, "You are also right."

Well, in this case, both the promoters of the Millennium Villages and the promoters of single, simple technologies are *right*. The lesson here is not to get caught up in doctrines.

8. Investment

In the second half of the 20th century, the government of The Netherlands decided to promote their greenhouse industry. To do this, it offered significant subsidies to farmers that constructed greenhouses based on strict government regulations. The Dutch greenhouse industry has been an exemplary one for many years since, bringing in billions of Euros to The Netherlands.

Not far from the town of Yako in Burkina Faso, there is a large artificial lake storing about 100 million m³ of water. The water from this lake was not utilized until scores of young people rented parcels of land around the lake, dug earth canals to move the lake water to fields adjacent to the lake and used plastic buckets to transport water from the canals to tomato fields.

The tomatoes are sold to Ghana traders that transport these all the way to Kumasi and Accra. This irrigated tomatoes-for-export industry around the lake was not created by governments or other organizations. It was a spontaneous development that resulted from availability of a market, from traders that were ready to purchase and transport the tomatoes and from accessibility to this market through a reasonable interstate road network.

Government subsidies of certain inputs such as fertilizers can be a driver for development. The return to the country that farmers achieve by using fertilizers more than compensates for the subsidy given for these fertilizers (Dugger, 2007). Tony Blair, Britain's former prime minister, and Prof. Jeffrey Sachs (Sachs, 2006) have been claiming for years that agricultural development in Africa will move only through heavy public and private investment in this sector. They are absolutely right.

Currently, Asia countries invest around 8–14% of their national budget in agriculture, while during the green revolution years this was as high as 15%. (Fan et al, 2010).

In 2008, countries in Africa invested only 4–5% of their national budget in agriculture. In practice, however, one should be very careful where the money is designated. Otherwise, it might follow the billions of dollars already wasted in Africa. Investment without expertise, organization and good governance will never succeed.

9. Availability of and access to markets

Transportation infrastructure is essential for the development of agriculture. Fortunately, most West African countries are connected by roads and some by rail.

These roads (paved and unpaved) allow trade of agricultural goods such as grains and vegetables within a country and between countries. Due to this road network, early season onions produced in the Aïr Mountains of Niger can be transported thousands of kilometers to Abidjan, the capital of Ivory Coast.

The population of the sixteen countries of West Africa is more than 320 million. Therefore, the regional market should be more of a focus for African farmers than overseas markets. People who are not aware of regional market opportunities place the blame for stagnation of African exports on the EU and the US, who are heavily subsidizing their farmers (Anderson, 2006). But even if Western countries were to lift their farm subsidies, Africa would still be unable to export grains, its major agricultural product, to these countries for the simple reason that Europe and the US have a huge relative advantage in grain production over Africa, with or without subsidies.

Every day, twenty-five trucks loaded with tomatoes cross the Burkina-Ghana border. 40% of tomatoes consumed in Ghana come from Burkina Faso

Most big cities of West Africa have access to the Atlantic Ocean. Unfortunately, however, the sea route for regional transportation of farm products in West Africa is not being utilized. This seems to be a worthwhile opportunity for development.

One last example on this topic. In 2002, I made a quick survey of vegetable prices in the markets of Niger. I found that tomato prices in the period June–October (the rainy season) were three times higher than prices in the period December–May. This spike in prices resulted from very poor fruit set of tomatoes during the rainy season. I then compared the performance of about fifteen supposedly heat tolerant tomato varieties during the rainy season and found out that Xina, a variety that was bred by ISRA (the Senegalese national agriculture research institute) gave a reasonably high yield during this period. However, because of many years of neglect, the quality and yield of the Xina variety was very poor. Following five cycles of selection and purification, we created a high yielding rainy season tomato variety with round and uniform fruit. We gave it the name "Icrixina"—the Xina of ICRISAT. In a very short period of time, Niger vegetable producers started growing Icrixina during the rainy season. Now, there is a good year-round supply of tomatoes in Niamey market. Recently an NGO found out that Icrixina performs very well in tropical Congo. Thinking of it, this should not be a surprise since Icrixina was bred for rainy season production and the climate in the rainy season is basically a tropical climate.

10. Relative advantage

Agriculture is an industry, and like other industries, it relies on relative advantage for economic success. In agriculture, factors such as climate, soil, availability of water, labor, land and efficient production systems are just a few determinants of relative advantage.

Let us start with a few examples:

- The US, Canada, Europe, South America and Australia have vast farming enterprises, plenty and reliable rains, government subsidies, credit availability and advanced machinery. All of these make the production of grain and pulses in these countries profitable. In contrast, an African owning 2–3 hectares of land, with unreliable rains and poor soils, no capital and no farm machinery does not have a relative advantage in grain production over farmers in the above-mentioned regions. A single African farmer produces one to two tons of grain per year, whereas an American farmer produces more than 200 tons. At this time, African governments lack the resources to import most of the grain and pulses they need as other countries including Japan and Israel do. Thus, the idea that Africa should stop producing its own grains and pulses and purchase them from others is not an option yet. In addition, most farmers in semi-arid Africa "earn their living" from rainfed production of grains and pulses. While working on improvement of such production to reduce the relative advantage gap with other countries, dry Africa should start looking for opportunities in *which it has* a relative advantage and expand in this direction.

- An example for success through utilization of relative advantage is the expansion of the horticulture-for-export industry in Kenya. The climate of the Kenya highlands is ideal for production of greenhouse roses and other flowers for export to Europe. Nairobi airport is a hub for many international airlines that transport flowers and vegetables to Europe. Also, in Kenya, labor is relatively cheap. All these relative advantages have created a flourishing horticultural products export industry that accounts for more than 40% of the agricultural GDP (Gross Domestic Product) of this country.

- Tropical coastal West Africa is growing and exporting cacao, rubber, cashews and mangoes. The climate and soils there are perfect for these crops and access to markets is easily navigable by sea.

- In the Sahel, the production of quality low cost meat represents a relative advantage opportunity. Nomadic herders with many millions of livestock roam over the vast empty grazing lands there. Production costs are low because these animals live only from grazing and are herded by family children. The construction *of modern slaughterhouses* combined with a short period of fattening prior to slaughtering could result in the establishment of a very substantial and profitable meat export industry.

The author in a young drip-irrigated date palm plantation intercropped with lettuce

Another opportunity that would capitalize on the hot Sahel climate is date cultivation. While this climate is not suited to growing many other crops, the date palm is an exception. Dates *need* and thrive in very hot climates. The Sahel, thus, offers a significant relative advantage for dates. In West Africa, a large and ready market already exists as date fruit is cherished by most of the population that is

Muslim. Indeed, quality date palm plantations rank among the most profitable agricultural enterprises everywhere in the world.

Realizing the potential of dates, in the year 2000, I started a program called "Dates for the Sahel" in collaboration with Dr. Saidou Koala and Mr. Saidou Abdoussalam of ICRISAT. From the very beginning, we realized that the availability of ample quality date plants at low cost was a precondition for success. To satisfy these parameters, tissue culture was chosen as the most advanced and least expensive technique for date plant propagation. We calculated that if we were to import tissue culture material to the Sahel, the cost of one plant before planting would be around $40.00. However, if we could establish a tissue culture laboratory in the region, the cost would decrease to $25.00 per plant. But more importantly, by building a date tissue culture laboratory in the Sahel, the date plants would become easily available and accessible.

Unfortunately, all our efforts to attract public or private funding to establish a date tissue culture laboratory at ICRISAT-Niger have so far been unsuccessful. However, over the last eleven years, we have been able to import several thousand tissue culture propagated date plants and sell them to various entrepreneurs in Niger, Burkina Faso and Mali. Many of these trees are now bearing fruit including the 2-hectare plantation that we established at the Sadore-ICRISAT research station in West Africa (see photo above).

The winter in the Sudano Sahel is typified by relatively mild temperatures with wide day/night amplitude. Days are bright and relative humidity is low. This climate is ideal for vegetable raising, and these vegetables can be exported by road or by sea to the "less fortunate" markets of tropical West Africa.

Expertise is also a very important relative advantage. Knowledge, however, is a rare commodity in dry Africa. The import, creation and dissemination of knowledge in dry Africa will give it a relative advantage (e.g. the greenhouse industries in Kenya and Ethiopia), and this is another precondition for successful agricultural development.

Yet simply having a potential relative advantage is only part of the picture. The conditions enabling it to materialize are also necessary. For example, regional onion exports from Niger could not have been managed without a reasonable regional road system to coastal cities of West Africa.

Chapter 3

Strategies for agricultural development

For any development activity to succeed, it must answer the following three questions:

- What to do?
- How to do it?
- Who is going to do it?

The first question is the most important question of the three, and it will be discussed in the next three chapters. The last two questions are cardinal for the success of a program.

Often, development experts do not have good answers to the "what to do" question. They solve this dilemma by turning the "means" ("how to do it") into the objective ("what to do"). Many development programs are thus centered on topics such as networking, use of GIS (Geographical Information Service), creation of data bases and teaching farmers how to make decisions.

While these are very important issues, they are there to assist development—not to replace it. Many projects end up with mountains of statistics and voluminous databases. These may make a good impression on readers and donors but not on farmers whose main interest is a "real" answer regarding the "what to do."

Warning—If someone not fluent in agricultural development wants to try to implement the solutions suggested in the following chapters, he/she is cautioned not to attempt it without the guidance of proven experts or failure will surely result. For instance, when my TV breaks down, I do not try to repair it by myself. Rather, I always call on a qualified technician.

An example:

In partnership with the Israeli Embassy in Senegal, we were able to demonstrate the feasibility of the African Market Garden (AMG) technology in this country. At least two organizations in Senegal then tried to install the AMG on a large scale without consulting Israeli Embassy experts. This resulted in total failure of their AMG projects and in the waste of valuable resources.

When people with good will wish to develop agriculture in a place that is new to them, their first inclination is "Let us see what exists there and improve on it." In the case of the Sudano Sahel, research institutes and other groups have spent significant resources and many years on improving traditional rainfed crops, but no one had thought of irrigated agriculture, which in my opinion is the central solution for development in dry Africa, just because *it was not yet being used on a large scale* when they first came to the region. Then a number of NGOs became immersed with irrigation of vegetables. But they tried to "improve" the traditional hand watering technology that is extremely inefficient rather than trying better alternatives for irrigation—again, because the alternatives *were not yet being used.*

One of the principal rules that I have adopted in my many years of agricultural development is: The technology, crop or variety that is *not there* may often be much more important than *what is there.*

When one wants to develop agriculture in a region new to him, he must take a comprehensive look at the whole range of factors that determine the approaches and technologies to be used. These factors include: market availability, the climate, the soil, the existing natural resources such as water, access to land (tenure), the educational status of the farmers, the social structure, labor availability, access to markets and access to credit in addition to the existing crops and production systems, their advantages and constraints.

After integrating all these factors, one should come up with solutions based on the following considerations:

- Relative advantage (most important.)

- Existing (elsewhere) crops and technologies that can fit into this new environment

- Efforts such as research needed to develop appropriate crops and technologies

- Efforts needed to introduce new technologies and/or crops or to improve the existing system

Years ago, I was asked to evaluate the landscape plans for the new campus of the Ben-Gurion University of the Negev in Beer Sheva. To my astonishment, I found out that the landscape architect planned to cover the entire campus with the *Tamarix aphyla* tree. The leaves of these trees are grey, and they always look "dirty" because they catch a lot of dust. Furthermore, due to its extensive root system, *Tamarix* "steals" the water from adjacent plants. When I asked him why he had chosen this particular tree, his answer was: "It grows everywhere in Beer Sheva. I know it is ugly but it is safe." Fortunately, I was able to convince the architect to change his plan after showing him other more suitable trees that were also thriving in this city.

An example of this type of integrated thinking:
A macro-analysis of the current agricultural situation in the semi-arid tropics of Africa reveals the following facts:

In dry Africa, agriculture is practiced by about 70–80% of the population. Each household cultivates between 0.5 to

3.0 hectares of land. The main crops are *rainfed* cereals and pulses that have a relatively low market value. Yields are negatively affected by low soil fertility and water deficiency. Most of the production is consumed by the household. This results in poverty, land degradation and food insecurity.

- One solution to this problem is dismantling the family-based system to allow the concentration of land. This would lead to a much more efficient grain production system that might be able to compete with the rest of the world and perhaps lower the prices of the rainfed products. A change of this nature took place in the US in the first half of the twentieth century, but there the fast developing service and industrial sectors were able to absorb the manpower released from the rural sector. A similar process is occurring today in China where the rural population is migrating to cities for better paid jobs. At present, dry Africa does not have industries and services that can absorb the huge number of people that would be released following a concentration of agricultural land. Therefore, such a step now would be disastrous.

- A more practical solution would be the intensification of agriculture options using high value crops and production of irrigated horticultural crops:

  1. Horticulture is labor intensive and in this aspect (manpower), the rural family has a relative advantage over an extensive farming operation. A small farm

using intensive agriculture produces much more value per hectare than a larger farm growing rainfed or even irrigated crops.

2. If irrigation is practiced, droughts are irrelevant.

3. Contrary to rainfed staple crops, high value crops can pay for investment in manure and fertilizers to improve soil fertility and boost yields, and on farms expansion using loans.

4. High value crops and the products of irrigated horticulture bring income to the producer, leading to the alleviation of poverty.

One of the best strategies with *fastest* results for rapid economic growth in underdeveloped countries and regions is the introduction of proven technologies and expertise from other places. This is how countries such as Japan, South Korea and more recently China have been able to attain leading positions in the global economy in a relatively short period of time.

In Chile, a non-profit corporation known as Fundacíon Chile was jointly created in 1976 by the Government of Chile and the US giant, ITT. This corporation initiated a vast salmon industry, a forest products industry, a wine industry, large scale production of a range of berries, abalone mollusks and more. It did this by introducing existing technologies **from elsewhere**, and when necessary, **hiring outside experts** for the implementation of these technologies.

In all cases, Chile made maximum use of its natural *relative advantages* such as clear lakes with cold waters for salmon production and good soils and climate suitable for the production of quality wines. In most cases, this organization introduced crops and technologies that did not exist in Chile beforehand. I got to know Fundacíon Chile intimately in 1986 when, at their request, I submitted a plan to establish a research and development complex in arid Chilean lands. I was yet another expert hired by this foundation to import know-how from abroad.

Unfortunately, in Africa, there are no groups such as Fundacíon Chile geared towards importation of technologies. The international and local agriculture research institutes deal mostly with development of new improved varieties of *existing* field crops.

In 2001, when I started working on crop diversification at ICRISAT, I noticed that most vegetable varieties produced in the region came from Europe, and many were not adapted to local conditions. Realizing that there are thousands of vegetable varieties around the globe, serving as a good starting point, I began selecting open pollinated vegetable varieties more suited to the local conditions. In a period of ten years, with relatively little investment, we were able to select and purify about fourteen vegetable varieties that were superior to other varieties produced in the region. After this first step that guaranteed better varieties to the growers within a short period of time was accomplished, we were able to begin the long tedious process of introducing tolerances, etc. through conventional breeding.

While importation of technologies and crop varieties is one avenue for success, one should not faithfully adhere to this strategy. Almost everywhere in Africa, you can find local varieties (Konni okra or ISV 128 cowpeas) and technologies that are as good as or even superior to those that were introduced. You can find underutilized indigenous crops, such as dolich (*Lablab purpureus*) or indigenous systems and technologies like the Gaya system and the zaï hole that need to be up-scaled.

The next chapter of this book will examine integrated approaches that offers a range of options for agricultural development.

Chapter 4

Technologies and crops for rainfed agricultural development

This chapter, probably for the first time, gives readers a range of options that could be used to develop the rainfed agriculture system in semi-arid Africa. I begin with the rainfed system because of its current dominance and its importance for food production in the semi-arid tropics of this region. Because this system is not working very effectively, much can and should be done towards its improvement.

Many technologies and crops described in this chapter were developed by the crops and systems diversification team at ICRISAT-Niger and are little known. Altogether, they provide a bank of options that could accelerate the development of rainfed agriculture in dry Africa.

Low soil fertility is a major problem of the rainfed system. Most of the soils of dry Africa are inherently unfertile. In addition, there is extensive land degradation. No, "land degradation" is too gentle a term—a more accurate description would be *land loss*. Land loss in the Sudano Sahel results mainly from soil erosion by water and by wind, from overexploitation of soil nutrients and from soil compaction by tillage, animals and natural causes. In some countries, such as Senegal and Mali, soil salinization is also a major factor in land loss (Diatta et al 2001, Valenza et al 2000).

Harmattan dust storm. Millions of tons of fertile top soil are lost each year

I could not find reliable statistics for the annual rate of land loss in the Sudano Sahel. Someone once told me that each year Niger is losing 5% of its soils. From my observations, I believe this to be true. Some figures do exist on the rate of spot soil loss from wind erosion. Buerkert et al (1996) measured absolute soil loss of 190 t/ha in one year on bare sandy plots.

People who care about this issue should visit the Sahelian countryside during a *harmattan*. There, first-hand, they will experience the trade winds blowing from the Sahara Desert that carry huge amounts of dust across West Africa. The dust, it is claimed, comes from the Sahara. Well, perhaps some of it originates there, *but much of the dust comes from the unprotected croplands of the Sahel.*

These winds cause not only health and transportation problems but also enormous crop and soil damage. If you check Wikipedia for information on harmattan, you may come away believing that the main concern is that the harmattan disrupts airplane flights.

Who cares about flight delays? The Sahel is losing its soil.

Before a harmattan, the color of the soil is brown, owing to the decomposition of surface organic matter. After a harmattan, the color of the sand has become golden yellow because all the fertile soil has been removed.

I also invite you to observe the effects of a torrential rain storm to the Sudano Sahel. Millions of tons of soil are removed by floodwaters, and no organization seems to be taking action to prevent this from occurring, including members of the UN Convention for Combating Desertification (UNCCD).

Established in 1994, UNCCD is the sole legally binding international agreement linking environment and development to sustainable land management.

The first meeting of the UNCCD country delegations took place in Rome in 1997. Since then the parties (194 in number) have convened many times in many places. So far, the UNCCD has had little impact on desertification control in Africa. The Hon. Arba Diallo, an ex-executive secretary of the UNCCD, is an exception. Mr. Diallo has spent much of his time in dry Africa attempting to convince governments and donors to invest in desertification control. A replacement for this dedicated man is urgently needed.

Clearing for crop production

Firewood prospecting

Human drivers for land degradation

Bare millet field following grazing of crops residues

Dust storm approaches

Soil erosion from water

Erosion after single rainstorm

Flash floods
This is what was left

Where has all the soil gone?

Without words

With great expectation, I attended one of the first meetings of this group in Paris. Even so, I wondered why the meeting was held in Paris of all places—as Paris becomes a "desert" only during July and August when most of its population migrates to the beachside. In this meeting, I did not hear a single concrete initiative to address the real issues of desertification. This experience stopped me from attending future UNCCD meetings.

Africa is an ancient continent and its soils have undergone a process of leaching of soluble elements. Most bases (calcium, magnesium, sodium, and to some degree, potassium and phosphorus) have been leached from the soil, leaving behind elements with low solubility, such as iron and aluminum. In these soils, the pH is low, and aluminum and manganese toxicity are prevalent. The cation-exchange capacity is low and is determined mainly by the organic matter content of the soil. In soils with low organic matter (most soils), nitrogen content is also very low.

Human-induced land degradation results from clearing of perennial vegetation to create space for cultivated land and from tree cutting to provide firewood. In the Sudano Sahel and in other semi-arid regions of sub-Saharan Africa, the production system is classified as agro-silvi-pastoral, meaning an integrated crop-trees-livestock system. After farmers harvest the annual crops, livestock are permitted to graze what is left in the fields. Many farmers harvest the straw and forage that remains for animal feeding, fuel and fencing. The soil, therefore, does not "benefit" from the fertility brought about by leaving crop residues, as is practiced in conservation agriculture around the world. As a result, towards the end of the dry season, the soil becomes totally bare, exposed to wind and water erosion. Lack of soil cover by trees or by plant residues results in accelerated erosion. Wind erosion is dominant in the sandy soils, whereas water erosion is dominant in the heavier soils. As most soil fertility is stored in the upper 2-cm soil layer, removal of this layer by wind and/or water significantly reduces its fertility.

A scientist working at ICRISAT-Niger has demonstrated that the amount of nutrients in the top 2-cm soil layer, which is removed annually by wind erosion, is equal to the nutrients added by the recommended application of 200 kg/ha of a complete fertilizer containing the main nutritional elements for plants: nitrogen (N), phosphorus (P) and potassium (K) (Buerkert et al, 2006).

Final product. About 60% of Niger agricultural lands are degraded

Scientists working in the Sahel noticed that sandy soil absorbs rain water like a sponge. There is hardly any runoff in this type of soil. Therefore, they have concluded that in sandy soils, there is no water erosion—that erosion of sandy soils is caused only by wind. *But they are mistaken.* In sandy soils, wind and rain "cooperate" in full harmony to erode the land. First, the wind removes the sand and patches of lateritic soil appear on the surface. A crust impermeable to water is produced on top of the laterite resulting in accelerated water runoff. This runoff "bites" into the sand and carries it to temporary and permanent river beds (see the photo of water-induced land degradation). This is an interesting example of cooperation between the destructive forces of nature.

The common solution to low soil fertility proposed by many (particularly, the giant fertilizer monopolies and now, the Alliance for a Green Revolution in Africa), is application of fertilizers. Unfortunately, applying fertilizers is not a very good idea if it is not accompanied by massive attempts to curb soil erosion problems. *And to date, there has been no concentrated attempt to do this.* Without control of soil erosion, fertilizers actually mask the effects of erosion. Farmers can keep adding fertilizer for a period of time, receiving reasonable yields until ultimately, they wake up one day to find that *the soil under their feet has literally disappeared.*

As a matter of fact, many farmers are already complaining that after they start adding fertilizer to the soil, they need to increase the amount each year to prevent yields from deteriorating. This is evidence that fertilizers alone are not the answer. In addition, fertilizers *are expensive* and becoming more so due to the increase in energy prices and the monopoly of the big fertilizer companies.

In 2012, the cost of 1 kg of an NPK fertilizer in Niamey, Niger was $0.67. The price is even higher in rural areas due to transportation costs from Niamey and to other transaction costs. The recommended rate of NPK application in the Sudano Sahel is 200kg/ha, costing the farmer $134/ha. In the same year, the farm gate price of 1 kg of millet was $0.45/kg. This means that farmers applying NPK must harvest an additional 300 kg/ha of millet (almost *doubling* common yields) just to cover the cost of the fertilizer. In rainy years, based on our experience the application of NPK can result in more than 600kg increase in yield. However in drought years (two out of five in some Sahelian regions), farmers will lose money from fertilizer application because yields that increase due to fertilizer application will be much lower than 300kg/ha or not at all.

Then there is the matter of "organic matter". Marenya and Barrett (2009) demonstrated that low soil organic matter commonly limits the yield response to mineral fertilizer application. In their study, the authors demonstrated that in degraded soils (most of the Sudano-Sahelian soils) the very low organic matter content limits the marginal productivity of fertilizer such *that it becomes unprofitable at prevailing prices.*

"Micro-dosing" is an innovative technology developed at ICRISAT-Niger. This technique drastically reduces the quantity of fertilizer application needed. Spot application of 20 kg/ha of NPK can produce *the same yield* as the broadcasting of 200 kg/ha of fertilizer. While this sounds like a worthy proposition, one of my graduate students (Ali, 2010) broadcasted 200 kg and micro-dosed 40 kg of diammonium phosphate (DAP) to a millet field. Both treatments were compared to a non-fertilized field which served as a control.

Results showed that net nutrient uptake was much higher in the micro-dosing treatment than in the broadcasting treatment **and in the control without fertilizer**. In other words, micro-dosing apparently is not a sustainable technique because, in the long term, it will impoverish the soil much faster than when fertilizers are not added. Those organizations now promoting micro-dosing as a means to support their fertilizer doctrines must be aware of this fact.

You cannot get something from nothing. Isn't that what the law of conservation of energy is all about?

"Keep Africa covered"— the agroforestry systems of the Sudano Sahel

Agroforestry is a very common practice in the semi-arid regions of the world. People from temperate countries are often unfamiliar with agroforestry systems and tend to underestimate their importance.

The agroforestry system, common in the Sudano Sahel, is a solution for soil erosion and for soil impoverishment since tree roots hold the soil and tree cover reduces wind velocity, traps dust and reduces the impact of rain drops that seal the soil surface. Trees recycle nutrients and litter from leaves and dead roots enrich the soil. The roots of many leguminous trees fix atmospheric nitrogen.

In its traditional application, African farmers *do not* plant trees, but they assist the regeneration of useful trees such as the *Vitellaria paradoxa* or shea nut (karité) and of the hardy *Fadherbia albida* by preventing their destruction by livestock.

Tree planting and/or grafting of native trees with improved varieties should be the next step for development of the rainfed agroforestry system of the Sudano Sahel.

1. Technologies for rainfed crop production

Farmer Managed Natural Regeneration (FMNR)

In the past, the tree cover in the traditional agroforestry system of the Sudano Sahel was denser than it is now. The degeneration of the agroforestry system is increasing the rate of land loss.

Many people, including myself, have blamed the destruction of agroforestry on rural poverty. But Tony Rinaudo has proven us wrong. Tony is an Australian who worked for an NGO called SIM (Service in Mission) in the Maradi region of Niger. He noticed that many trees that had been destroyed could regenerate from stumps, roots or seeds. The problem, however, was that the young plants were either eaten by livestock or "weeded" out by farmers.

New trees are sprouting under FMNR

In the early 1990s, after giving much thought to this "sustainability" aspect, Tony developed a technique that he called "Farmers Managed Natural Regeneration" (FMNR). In this system, farmers protect the young tree saplings that appear spontaneously in the field, prune them and shape them in order to get maximum firewood production.

The Bull-Sheet Technology

In the Sahel, one can distinguish between primary (farmers) and secondary (nomads with herds) land users. Each of these two groups believes that they have the right to the same lands. After farmers harvest their crops, the nomads bring their herds to graze the aftermath. The nomads do not like tree planting because they see it as a threat to their grazing rights. Thus, many times nomads make sure that any trees planted by farmers are eaten by their herds. I sought a solution to this problem for a long period of time. One day, I consulted Prof. Hamidou Boly the ex-director of INERA, a friend and a gifted veterinarian. His suggestion: "You should coat the young trees with animal dung. This is an excellent repellent. Livestock will not touch a tree coated with dung."

Following his advice, I produced a spray composed of cow dung mixed with water soluble plastic glue that allows the attachment of the dung to the young plant. I named this technology—the "Bull-Sheet" Technology.

And it works. Young planted trees should be sprayed once a month with cow dung. This is a simple solution for a serious problem.

Tony brought this technique to farmers who accepted it and started spreading it with the support of government and other organizations and on their own. Now about twenty years later, five million hectares of land in the Maradi-Zinder region of Niger are covered by regenerated trees.

Tony was interested in quantifying the economic benefits of FMNR. Through an Australian friend who worked at Bioversity International (yet another CGIAR Institute), he applied for a Mickey Leland International Hunger Fellowship for a US researcher named Eric Haglund to conduct this study. He then asked me if Eric could be hosted and assisted by ICRISAT-Niger, which we gladly did. Since I am not an economist, I asked Dr. Jupiter Ndjeunga, ICRISAT's excellent agro-economist to supervise Eric's work and he agreed.

Two years later, Eric was able to show that the income per capita in farmers' households practicing FMNR was 57% higher than the income of non-practicing farmers. He further demonstrated that this higher income was mostly due to the increase in crop yield (Haglund et al, 2011). The higher yields were likely the main reason for adoption of this technique by farmers—not the extra firewood as some claim—even though it was necessary to wait several years until the trees were large enough to materialize their effect on yields. Due to an oversight on my part, Eric did not measure soil fertility.

Spraying young Acacia trees with cow dung using a wall plastering machine

I can only guess that the increase in crop yield resulted from less soil erosion combined with the addition of organic matter by the trees leading to improved soil fertility, as has been demonstrated in other agroforestry systems (Breman and Kessler, 1997).

Many organizations are now spreading the "gospel" of FMNR in other countries of dry West Africa. Chris Reij, a natural resource management specialist, is heading this new movement for re-greening of the Sahel with FMNR. I wish Chris and others success in this undertaking.

I see the "FMNR movement" as a starting point for introducing useful *planted* trees to the drylands system and for raising the potential of wild trees through grafting with improved varieties.

Fadherbia albida agro-forestry

In his comprehensive study of FMNR, Eric assessed the preferences of farmers for the trees being regenerated in their fields. *Fadherbia albida* came first on the list. It also was the top preference in another survey carried out by ICRAF (2009). *Fadherbia albida* (also known as *Acacia albida*) is a leguminous tree of the dry regions of Africa and the Middle East. In dry Africa, it is a long living tree that can grow to a height of 15 meters. *F. albida*, contrary to most other trees of dry Africa, *sheds its leaves at the beginning of the rainy season and regains them during the dry season.*

Because of this unusual behavior, during the rainy season, *F. albida* does not compete with annuals planted under it for light, water and nutrients. Thus, it is an ideal agroforestry species.

In poor soils, millet grows only next to *Fadherbia albida* trees

Lush millet stand in a dense *Fadherbia albida* forest

This tree has other very important advantages. Being a legume, its roots produce symbiosis with bacteria called *Rhizobium* that fix nitrogen from the air making it available to the plants, nitrogen being the most important nutrient for plants. The litter that it sheds and dead roots further enriches the soil with organic matter and nutrients. Apparently, the roots of this tree bind the soil, thus, minimizing erosion (see the sand mound by the *F. albida* tree top left), and it's very deep roots recycle leached nutrients. A dense stand of F. albida also reduces soil erosion by acting as a wind break. The pods are eaten by ruminants and are sold at high prices in rural markets and so are its leaves. Furthermore, its branches are used for firewood. It is no wonder that farmers practicing agroforestry prefer this tree species above all others.

Dr. Dennis Garrity, one of the previous DGs of ICRAF (the World Agroforestry Center) is a big promoter of *F. albida* agroforestry. When I attended the Gates Foundation Forum for the New Green Revolution for Africa in Accra on September 2010, to my surprise I noticed that Dennis was standing near the registration desk with a stack of pamphlets advertising *Fadherbia albida* agroforestry. There were two reasons for my surprise. The first was that usually the director of a research institute does not distribute pamphlets. This task is delegated to someone on his administrative staff. The other reason was that this man was trying to convince the Gates Foundation representatives attending this meeting, who were great promoters of fertilizer application, to use a "substitute" for their favorite fertilizers. This action required a lot of courage from a person in such a central position. Dennis, I take my hat off to you. Africa needs more people like you.

According to the belief system of the Serer people of Senegal, *F. albida* is considered a holy tree that must not be cut down. This tradition has resulted in dense populations of these trees in their fields. Whenever one comes across a population of closely crowded *F. albida* trees, one instinctively knows that these fields are fertile and that they will produce high yields of annual crops. Moreover, research conducted in Zambia demonstrated that the amount of organic matter and of nitrogen under *F. albida* trees in a field planted with maize was **double** the amount in the soil of maize fields without *F. albida*. When fertilizers were not given, the grain yield of maize growing under the trees ranged from 5.1–5.6 tons/ha. This is the kind of maize yield that one gets in Zambia with *ample fertilizer* application. At the same time, maize growing *without F. albida* trees and *without* fertilizers produced a yield of 2.6 tons/ha— less than half the yield of maize planted under the trees.

Apparently *Fadherbia albida* planting is the best known solution for Sustainable Land Management (SLM) in semi-arid Africa.

The solution was there all the time but we did not take notice of it.

Note: FMNR is actually becoming a "competitor" of *Fadherbia albida* agroforestry because of the intensive drive of many to generate resources for large-scale implementation of FMNR. However, *planting F. albida stands gives significantly better yield results than FMNR.*

Alley farming

Yet another agroforestry system is alley farming, where rows of perennials are planted at intervals in a field of annual crops. Its introduction has been attempted by many scientists, myself among them (Pasternak et al, 2005). The theory behind this technique is that perennials are supposed to increase soil fertility by nitrogen fixation and by adding organic matter to the soil mostly through the spreading of the mulch produced by the perennials in the alley. For a range of reasons which are unclear, most attempts to introduce alley farming, including my own, have been unsuccessful.

Based on the lessons learned from FMNR, where farmers have demonstrated the patience to wait four to five years until the trees begin to affect soil fertility and crop yields, it appears to me that we can convince farmers to actually plant and protect *Fadherbia albida* in their fields. The density of 100 trees/ha is recommended by ICRAF (and by the Government of Zambia). If the governments of dry Africa, **backed by the UNCCD,** would adopt mass planting of this species as a national challenge, **many of the land degradation problems of dry Africa could be resolved.**

I am not the first to claim: "Keep Africa covered." Please notice I did not say, "Keep Africa green," as many people do. The issue is not about color; it is about the prevention of the massive loss of soil and its enrichment with nutrients using *fertilizer alternatives* i.e., trees and shrubs.

Bioreclamation of Degraded Lands (BDL)

Every time that I travel in the Sahel, I come across degraded lands composed of crusted lateritic and/or compacted clay soils. This depressing reality is more prevalent in Niger, where to my estimate more than 60% of the land is degraded. What is the reason behind this disastrous land degradation?

The problem actually started long, long ago in geological times when the climate of the Sudano Sahel alternated between humid tropics and dry tropics. When a humid tropical climate prevailed, particularly in the Paleo-Eocene age (55 million years ago), lateritic soils were formed. When the humid tropics were replaced by long, dry periods, these soils were covered over by sand, and then, in time, the tropical climate returned. If you dig into the Sahelian soil, you will find alternating layers of laterite and sand, reflecting climatic changes that have occurred over the ages.

Starting in the Neolithic age, droughts and land clearing for crop production have resulted in the destruction of the tree cover. In the rainy season, farmers have continued to plant millet, cowpeas and (more recently) groundnuts in the sandy soils. In the dry season, crop residues were grazed by livestock, leaving the soil totally bare and subject to wind erosion. Over time, the wind has removed the sandy layer, and the latest lateritic layer has been exposed. When covered by sand or vegetation, the laterite is a porous soil with a reasonable infiltration rate. However, when exposed to intermittent wetness and dryness at high temperatures, an impermeable crust forms on the top, preventing water infiltration. The result is classical "desertification." It is common to find crusted, degraded lands almost totally bare of vegetation in regions even with annual rainfall of 800 mm. A "desert" has been created although rainfall may be greater than that in many zones that carry a lush vegetation.

Over the years, there have been attempts by Sahelian governments and others to reclaim the bare degraded lands. The main activity in this direction has been the digging of micro-catchments that resemble half-moons, and they are therefore called *demilunes* in French. *Acacia senegal,* a hardy small tree that produces the famous gum arabic, was planted in these demilunes. A problem occurred because the land and the acacia trees belonged to the community. Since the land had no private ownership, the young trees were destroyed by grazing animals. I have no estimation of the area planted and subsequently destroyed and replanted, other than it is extensive.

Mamadou Tanja, a former president of Niger, was very fond of *A. senegal* plantations. Every year, he promoted their planting particularly on "Tree Planting Day," which coincides with Independence Day in Niger. In 2002, about 18 million of these trees were planted on communal lands. Unfortunately, they disappeared within a few months, mostly in the stomachs of goats and other livestock. In a very few places, trees did survive and produced thick stands—but no one is harvesting the gum Arabic because the low gum yield of the wild *A. senegal* discourages commercial harvesting. At the very least, soil fertility is being restored, and the Sahel "looks green."

Degraded lands are often reclaimed by governments with heavy machinery, producing earth bunds that collect rain water. Earth bunds are constructed over large areas, and then nothing further is done with these areas. True, water collection results in more grass production around the trenches which can feed animals, but this contribution is negligible.

Another technology introduced by some NGOs to restore degraded lands is the zaï hole planted with sorghum. Although this system is effective, farmers in many regions of the Sahel have been slow in adopting it because of the great effort required to dig these holes in crusted soils. However, I believe that adoption of the zaï technology will happen because with the coming climate change, zaï holes which harvest rain water, will help mitigate the negative effect of long dry spells between rains associated with this change.

Lateritic soils are actually **more fertile than sandy soils**; both water holding and cation exchange capacity are greater than that of sandy soils. *It is the crust that renders them into degraded lands.* One can further distinguish between "hard" laterites and "soft" laterites. The hard laterites are composed of thick gravel and small stones and are very difficult to cultivate. These soils are suitable for dryland tree plantations using micro-catchments (demilunes) to collect rain water for the trees. In contrast, soft lateritic soils are composed of fine gravel, sand and silt. They are easier to cultivate and have a higher water holding and cation exchange capacity than hard laterites. Soft laterites can be used for production of annual crops as well as for trees. In Mali, farmers call the hard laterites, male laterites and the soft laterites, female laterites.

After extensive observation and analysis of non-sustainable activities to reclaim degraded lands, and after much thought, we developed the Bio-reclamation of Degraded Lands (BDL) approach for reclamation of degraded lands that provides answers to most of the constraints, such as communal lands and planting of low value crops, which have led to previous failures (Pasternak et al, 2009). The principles of the BDL are as follows:

- Reclamation using the BDL system should concentrate on soft laterites or on less compact clay soils.

- The degraded lands to be reclaimed should be owned by farmers

- Women should be responsible for the cultivation of these reclaimed degraded lands.

- High value horticultural crops should be planted in the degraded lands with emphasis on crops that are traditionally cultivated by women. The high value obtained from these crops would justify the labor necessary to construct and maintain the various water harvesting structures.

In West Africa, although women are not allowed to own croplands, they are permitted to cultivate degraded lands. The degraded lands usually belong to the community and are managed by the village chief. Therefore, it is possible for the chief to allocate these lands to a group of women for extended periods of time. The womens' group distributes the land among its members and protects individual women from outsiders who may wish to take control of their plots. The optimal size of an individual womens' plot should be around 400 m² for best management. The net value from this plot after trees begin to bear fruit is expected to be around $150 per year—half the annual income of farmers in many Sahelian countries. Income from the BDL can be even higher if partial irrigation increases yields of fruit trees, of moringa and of vegetables, while allowing planting of other drought tolerant vegetables such as pumpkin and watermelons.

For the mildly undulating slopes prevalent in the Sudano Sahel, BDL guidelines recommend three water harvesting techniques:

- Micro-catchments (demilunes) planted with Pomme du Sahel, the domesticated *Ziziphus mauritiana,* at a spacing of 10 x 10 meters. The dimensions of a demilune should be 3 x 3 meters.

- Zaï holes planted with annual hardy vegetables such as okra, roselle and cowpeas set up at a 0.5 x 0.5 spacing. The zaï dimensions should be 20 cm wide and 20 cm deep.

Harvesting okra and roselle in a BDL in a year of 250 mm rain

- Tied trenches dug every 10 meters, perpendicular to the slope to catch the water not collected by the zaï holes and demilunes. *Moringa oleifera* and

- Henna (*Lawsonia inermis*) could be planted in the trenches at a spacing of 4 meters between trees.

Demilune and zaï holes in the BDL for water harvesting

The BDL technology offers many solutions to the cardinal problems of the Sudano Sahel; empowerment of women, income generation, improved nutrition, land reclamation and combating climate change are among its benefits. This technology evolved over a five-year period during which various sites were established in many regions of Niger and Senegal. So far about 700 BDL sites have been established in these two countries approximately benefiting 50,000 women plus their families.

If one does not plant trees in the BDL there is no need for the expensive fences to protect the perennial plants against grazing animals. Women plant the traditional vegetables in zaï holes only during the rainy season when grazing animals are forbidden to roam in agricultural fields.

BDL technology is now ready for mass dissemination. Although it will encompass a very small proportion of the degraded soils in the Sudano Sahel, for those employing it (meaning women and their families), it offers positive life-changing aspects.

High value tree crops for Ethiopia's drylands

The highlands of north Ethiopia are basically an extension of the Sahel. It has a semi- arid climate and rains fall mainly from June to September with a short rainy season in March. The first time that I visited the mountainous region of Tigray, it struck me how similar this region was to the Galilee of Israel. However, something was missing. After a while, I realized what it was. While the Galilee is covered by olives, grape vines, figs, pomegranates, oaks, carob and pine trees, the landscape of the Tigray mountains was totally barren.

Over thousands of years, farmers of the dry Mediterranean basin, of dry Central Asia and of Mexico have developed an agroforestry system based on fruit trees— a system that is still sustainable in our day. In the Mediterranean basin farmers plant olive and other dry land fruit trees at a wide spacing, and in the rainy season, they intercrop the trees with wheat, barley, chickpeas, sesame and other crops. The products from the trees bring a higher income than the products of annuals alone, particularly during dry years when the more drought tolerant trees can continue to yield while annual crops wilt away.

For a reason beyond my understanding, Ethiopia with its thousands of years of agricultural tradition has never developed fruit tree-based agroforestry. Perhaps this is because smallholders did not own the land. I then started a small project to introduce hardy Mediterranean fruit trees to this region, calling it "Trees from the Holy Land" to capitalize on the great affinity of Ethiopians for the Holy Land. I brought in many varieties of olives, grapes, pomegranates, figs, almonds and Pomme du Sahel (the one species that did not originate in the Mediterranean or Central Asia regions) to Tigray.

Dr. Debesai Senbeto worked in Tigray for an NGO called REST (Relief Society of Tigray) as the director of the forestry department. This outstanding person stood behind the reclamation of a huge amount of bare degraded land and planted all of it with trees. Dr. Senbeto agreed to cooperate in the "Trees from the Holy Land project."

In six years, Dr. Senbeto recuperated and planted 300,000 hectares of forest in the steep mountains of semi-arid Tigray

Although the olives grew very fast, a great disappointment was that they were late in setting fruit at elevations below 2,900 meters. However, all the other species did extremely well. The "Trees from the Holy Land" project has been transferred to World Vision–Ethiopia and has the support of Tony Rinaudo, who is now working for this organization.

Dr. Debesai Senbeto in one of his tree nurseries—a giant in size and spirit

In the semi-arid regions of Ethiopia, the production system will include terraces, trenches, earth and stone bunds, micro-catchments and zaï holes to maximize water harvesting and minimize land degradation. I do hope that someday someone in Ethiopia will wake up to the potential of drought tolerant trees for their vast semiarid mountainous regions.

In the meantime many development agencies are distributing money to Ethiopian farmers who cannot make a living from their lands.

2. Rainfed tree crops

Hardy shrubs and trees are much better adapted to harsh growing conditions than are annual herbaceous plants to which annual crops belong. Trees are able to thrive in non-fertile soils and grow in environments exposed to frequent droughts and extreme heat. There are several reasons for this:

- First, hardy trees have a more extensive root system than annual crops that permit the exploitation of water from a larger volume of soil. Moreover, leaves are coated with a waxy cuticle, much thicker than that of the annuals that prevents water loss. For these reasons, trees can survive "dry spells" (long dry intervals between rains) better than annuals. Some dry region trees even shed their leaves and become dormant during the dry season, requiring very little water during this period.

- Another factor giving trees an advantage over annual crops is that their root systems are able to reach leached nutrients in the soil. In addition, many dryland trees are better adapted to high temperatures and low humidity than are annuals, which usually grow during the cooler and moist rainy season.

- But perhaps the most important consideration of all is that tree roots enrich the soils in which they are planted and reduce soil erosion (one of the greatest threats to dryland agriculture), while adding nutrients to the soil. This feature is particularly notable in leguminous trees that their roots fix nitrogen from the air.

Currently, trees and shrubs play a more significant role in agriculture in tropical African regions than in dry Africa. There, bananas and cassava are the main source of carbohydrates, while trees such as mango, cacao, coffee, palm oil, coconut and cashew provide raw materials for processing and exports. In contrast, there are few tree species of value in dry Africa at present. This imbalance, however, can and should be changed through research and development and through the creation of agro-industries that process the products of dryland trees.

One of the best ways to promote plantations of drought tolerant fruit trees in Africa's dry regions is the establishment of local juice industries that use the fruit of these trees. The international juice market is a multi-billion one and is growing at a rate of 5% per year. Many juice ingredients consumed in developed countries such as mangoes, citrus and pineapples are produced in tropical African countries and exported as concentrates to developed countries.

Mangoes for juice

The Dafani juice factory in Burkina Faso is a shining example of what can and needs to be done to a greater extent in dry Africa. Dafani built a modern juice processing factory in the region of Banfora, which receives about 1,000 mm of rainfall per year. This region is heavily planted with mango trees and until the factory was established, there were few market outlets for the fruit. In a period of only two years, Dafani was already supplying 70% of the juice market of Burkina Faso and had started exporting to neighboring countries such as Niger and Mali. Dafani owners aim at purchasing mangoes at reasonable prices from 20,000 small mango producers in the region.

Dafani, modern juice plant—purchasing mangoes from 20,000 small producers in Banfora, Burkina Faso.

Marula for juice and oil

Marula (*Sclerocarya birrea*) is a native of semi-arid Africa. It is highly tolerant to drought and can be found in regions of 200–300 mm/y rainfall. This tree is relatively fast growing. It loses its leaves during the dry season, and north of the equator, it produces fruit in the months of June and July at the beginning of the rainy season.

In the vicinity of the town of Maradi in Niger, there are natural marula forests stretching over thousands of hectares. An excellent feature of marula (*Dania* in the Haoussa language) is that fruit falls to the ground when half ripe and can be collected from the ground, transported to the factory and allowed to ripen there.

The juice produced from marula fruit is very tasty and highly nutritious. Research in Israel has demonstrated that it contains more anti-oxidants than the juice of pomegranates and can effectively reduce blood pressure in people with hypertension (Borochov-Neori et al, 2008). The seeds of marula contain quality oil with applications in the cosmetics industry.

I started promoting the establishment of a marula factory in Maradi with the idea that it could purchase the fruit from women and children who would collect it in the wild. This will result in a new source of income for women. I have also planted marula trees inside micro-catchments in degraded lands with excellent results. The establishment of marula juice factories should promote large-scale planting of this species on degraded lands of the Sahel. This is a dream that could be fullfiled if only the development agencies or private sector had the vision and courage to do it. Can you Imagine? All these degraded lands covered with Dania for juice that is exported to Europe and the US.

Saba senegalensis for juice

Saba senegalensis is a vine (liana) of the Sudano region of West Africa. The saba fruit is the size and shape of a mango fruit. West Africans traditionally produce a sweet and sour juice from it. At present, saba is picked in the wild and sold in most city and village markets of the Sudano Sahel.

Dr. Albert Nikiema, who worked on the crops and systems diversification team at ICRISAT as a tree and nursery expert, demonstrated that saba can be grown on trellises as a plantation crop and if irrigated, it will produce fruit for eight months of the year. He also demonstrated that there is very high variability in fruit yield and quality among the seven provenances that he has tested and among plants within each particular provenance, thus, allowing the selection of superior lines.

Recently, the Dafani juice company started mixing mango juice with saba juice. This product is considered by many to be even tastier and healthier than pure mango juice.

The story of Pomme du Sahel (PDS)

Ziziphus mauritiana is a small tree native to semi-arid Africa and South Asia. The French name for this tree is *Jujubier*.

It produces a small edible fruit that is sold in Sahelian markets in its dry form.

Z. mauritiana is drought, salt and flood tolerant. It has a deep root system that allows it to draw water from great depths. It loses its leaves during the dry season and sprouts before the beginning of the rainy season. Like many other *Ziziphus* species, the tree has a long life.

This tree was domesticated in India where it is called ber. The domesticated ber has a fruit 10–30 times larger than the fruit of the wild *Z. mauritiana,* and the fruit of the domesticated tree is very tasty and highly nutritious. The fruit of the domesticated ber looks like a small apple but (in my opinion) it tastes better. *Nutritionally, it has three times as much protein, four times the calcium and fifteen times as much iron and Vitamin C as apple..*

Pomme du Sahel fruit (left) next to fruit of the native Z. *mauritiana* (right)

The domesticated *Z. mauritiana* was first introduced to the Sahel in 1996 by Dr. Modibo Sidibe, a researcher from the local NAR called IER (Institut d'Economie Rurale) in Mali.

In 1998, as the director of IPALAC, I sponsored the visit of Dr. Brij Vashishta (the scientist who popularized the domesticated ber tree in India), from CAZRI (the Central Arid Zone Research Institute) in Rajasthan in northwest India to Mali to train people in the propagation and utilization of ber. At that time, I sent my technician to graft some wild *Z. mauritiana* trees with five varieties of the domesticated ber in the yard of ICRAF at the Samanko site near Bamako.

Two varieties of Pomme du Sahel

Another mechanism of dissemination of the domesticated ber plants was through agro foresters from Senegal and Burkina Faso who attended IPALAC's "Trees for Arid Lands" annual workshop in Beer Sheva, Israel. At the end of the workshop, these scientists were given plants to take home with them. These dissemination mechanisms started the dispersal of the domesticated ber in three (Senegal, Mali and Burkina Faso) Sahelian countries. The names of the varieties supplied were: Kaitali, Umran, Seb, Gola and Ben-

Gurion. The latter is a variety selected by the team of Prof. Yossi Mizrahi, the arid lands fruit tree expert at Ben-Gurion University and is named after the university. It is high yielding and the fruit is very tasty.

Grafted Pomme du Sahel plants at ICRISAT nursery—70,000 plants were sold each year

The first thing that I did after arriving at Sadore, the experimental station of ICRISAT-Niger, was to graft the five domesticated ber varieties on mature local *Z. mauritiana* trees growing at this site. The Pomme du Sahel varieties used had been introduced to Burkina Faso by Dr. Albert Nikiema, the director of CNSF (Centre National de Semences Forestieres), one of the scientists who had attended our workshop in Israel and who received some of the *Ziziphus* plants prior to departure to his country. Dr. Nikiema sent his top tree grafter with scions of the five varieties to Sadore. The grafting was completed in one week.

When I started the promotion of the domesticated *Ziziphus mauritiana* in Niger, it was brought to my attention that the fruit of the wild *Z. mauritiana* suffered from a bad reputation because it was perceived that it is mostly eaten by small children and by goats. So my first task was to change its image.

I decided to call it the "Apple of the Sahel" or in French, *Pomme du Sahel* (PDS). The fruit resembles a small apple and its taste and texture are also similar. Now, Sahelian residents could proudly say: "Not only the Europeans have apples, but so do we."

Our first step in the dissemination process of this species was to build a nursery to propagate it. Against many objections from my local staff, I trained a group of women (rather than men) to graft the trees, and from 2004 until 2012, this nursery has been producing about 70,000 grafted PDS trees per year.

We wanted to promote the fruit in Niamey, but we did not know how to accomplish this. However, as it often happens, salvation came from an unexpected quarter. We started selling fruit from our various trials to the ICRISAT staff living in Niamey, the capital of Niger. The staff gave the fruit to their children to take to school. These children distributed the fruit to other children at school who took it home for their parents to taste, helping the fruit to gain immediate popularity. We then advertised on the local radio and in newspapers that we were selling the plants at the ICRISAT office in Niamey. In a short period of time, a big demand for PDS was created in the capital.

Since 2006, the ICRISAT nursery has sold hundreds of thousands of PDS trees to residents of this city, who have planted the trees in their backyards. In addition, plants have been sold to private fruit tree growers all over the country. At this stage, we started training village nurserymen and women to graft the plant, yet another venue for dissemination. Pomme du Sahel became very popular, and ICRISAT's connection to it was very well known. For instance, whenever I was stopped in Niamey by the traffic police, I would tell them that I was from ICRISAT. "ICRISAT," they would say, "Ahhh Pomme du Sahel." And they would let me go without a fine. Well, this act they thought was their contribution to development of their country.

Training youth on grafting wild *Ziziphus* trees with Pomme du Sahel

There are two problems with planting young trees, particularly in rural areas. The first is that livestock can destroy the young saplings, and the second is that it takes about six years for the Pomme du Sahel tree to reach full maturity even though its first fruit is produced after the first year. Many times farmers do not have the patience to wait this long for the trees to produce a reasonable quantity of fruit.

These problems have been successfully addressed by grafting existing mature wild trees. As I already mentioned, when I began working at ICRISAT in 2001, my first action was to graft mature trees of wild *Ziziphus* with improved Pomme du Sahel (PDS) varieties. For some reason, I never thought that this same technology could also be applied in farmers' fields where the trees grow wild. Then one day in 2009, one of my technicians took the initiative and grafted mature *Ziziphus* trees growing in the field of one of our pilot farmers. After grafting, the trees gave fruit in just five months. Needless to say, the farmer was thrilled.

Oualam and Tera are two of the poorest and most drought prone regions in Niger. These regions however have plenty of wild *Ziziphus* trees. A year after our grafting success, I brought young people from Oualam and Tera to ICRISAT's Sadore station for training in grafting the mature wild *Ziziphus* trees with Pomme du Sahel. The teens (ages 16–18) received five days of training on tree grafting and tree care. At the end of the training, they each received a grafting kit. When they returned home, the young people started grafting the wild trees in farmers' fields, receiving payment from farmers for their work.

Farmer inspects his grafted PDS tree, five months after grafting. Note the millet stalk fence around the tree.

I visited the region of Tera five months after the first trees had been grafted, just before fruit ripening. I brought some mature PDS fruit with me and distributed them to farmers. After tasting the fruit, one of the farmers commented: "Prof. Dov, this is fantastic. Just one fruit can provide breakfast for each of my young kids. I will now take double care of my grafted trees that are ready to give fruit." Can you imagine—one PDS fruit for a child's breakfast in poor Tera? At least the children will get more vitamins and minerals from the fruit than from their traditional breakfast.

Rain-fed PDS trees offer several advantages: A mature tree will bear a minimum of 30 kg of fruit per year. At present, 1.0 kg sells for $1.0 meaning income from a single tree is about $30. Five grafted trees will thus provide the same gross revenue as one hectare of millet planted in these poor marginal regions. And the nutritional value of the fruit is superior to that of millet. No less important is the fact that PDS trees will bear fruit during drought years when yields of annual crops fail. There is ample opportunity for grafting hundreds of thousands or even millions of wild *Ziziphus* in the Sudano Sahel with scions of selected PDS varieties. It is such a simple and effective means with which to improve food security, nutrition and income.

Actually, we have initiated a massive tree grafting project in East Senegal through a USAID food security program called Yaajeende. In 2012, about 250 young people were trained in grafting large wild *Z. mauritiana* trees with quality PDS varieties. In the first year, a total of 9,000 trees were grafted. We hope that 100,000 wild *jujubier* will be grafted before the project ends. This translates to 2,000 tons of fruit that will be available—using a technique that is so simple and so effective.

FMNR (page 38) farmers in Maradi-Niger are now hiring grafters trained by ICRISAT to graft the wild *Ziziphus mauritiana* trees that sprouted spontaneously in their fields.

Research at ICRISAT-Niger demonstrated that rain-fed Pomme du Sahel plantations gave a profit ten times higher than the profit from millet-cowpeas intercrop ($500 versus $93 per hectare).

When Pomme du Sahel trees were intercropped with watermelons annual profits climbed to $1,370/ha.

Sweet Tamarind

Tamarind (*Tamarindus indica*) grows wild in the Sudano Sahel, and its pods are used in juice production as well as a sauce in traditional dishes. The most difficult logistical task of our tamarind tree introduction activity was the acquisition of sweet varieties of this fruit. Thailand is the origin of the sweet tamarind and exports large quantities of fruit and juice. After identifying a source of scions in Thailand, we inquired about the consequences for exportation of sweet tamarind scions to Africa. The answer came a week later: "The consequences will be five years in jail for the person who attempts to export them."

Well, this door was apparently closed to us. But we were determined to continue our quest. We discovered that in Florida, researchers from the USDA (United States Department of Agriculture) have a collection of four sweet tamarind varieties that came from Thailand through the Philippines. They readily sent us the scions by DHL (with all the health certificates). However, the DHL dispatcher was certain that there was no such country called Niger. He/she "corrected the mistake" by sending the scions to Nigeria. Thus, it took three weeks for the scions to finally reach us. Surprisingly, the grafting of three out of four varieties was successful—quite a hardy plant it is.

I would like to use this opportunity to praise the US government for not only permitting but actually encouraging not-for-profit entities such as research organizations to freely use all the germplasm (plant material) that is at its disposal. Over the years, we have received sweet tamarind scions, fig cuttings and okra seeds from their central germplasm banks, all free of charge and—free from fear of imprisonment by greedy governments and free from punishment by executors of Article 15 from the controversial UN Convention on Biological Diversity which the US refused to retify. This generosity and openness results in huge benefits to the poor farmers of the developing world and to the economies of poor countries. USDA, please keep up your good work.

Karité (*Vitellaria paradoxa*)

Vitellaria paradoxa, called shea nut in English and *karité* in French, is a very important tree in the agroforestry systems of the Sudano region due to its many uses. From the kernels of the fruit, women extract a butter-like substance that is used as cooking oil and in various cosmetic products and for production of traditional soap. Shea butter is also used as a cocoa substitute in the chocolate industry. In recent years, the international cosmetics industry has begun using shea nut oil and is now paying high prices for it.

This is an opportune time to domesticate this species, and indeed individual trees with large kernels have been identified by various organizations. In Burkina Faso, scions of improved karité have been grafted onto mature wild trees, and it is expected that the newly grafted trees will give at least double the yield that they gave before.

Acacia tumida on degraded lands for renewable firewood

Acacia tumida is a native of semi-arid Australia. It normally grows in sandy soils and thrives in both basic and acidic soils. Out of eight drought tolerant tree species that were planted in micro-catchments in a degraded, hard lateritic soil, A. *tumida* performed best. Its seeds are rich in protein, oil, fiber and micro-elements and can be used for chicken feed. Average annual seed production in degraded land in a region with annual rainfall of 500 mm is around 200 kg/ha.

Two-year-old *Acacia tumida* planted in demilunes on degraded land—exceptional biomass production

Unlike many other Australian acacias such as *A. colei* and *A. cyanophilla, A. tumida* is a long living species. If pruned at a height of one meter, it regenerates well. For this reason, I call it a tree for "renewable firewood." If pruned every two years, this tree can produce one ton of firewood per hectare per year in a region with 500 mm/y rainfall.

Plantations of *Acacia tumida* on degraded lands in the vicinity of Sudano Sahelian cities may be a solution for the current destruction of woody species that supply the ever increasing demand for firewood. However, as aforementioned, planting *must not* be on government or communal plantations. When something belongs to everybody, it actually belongs to nobody. Trees planted in communal lands will be easily destroyed by livestock or cut for firewood by intruders. The plantations must be owned by farmers who will care for and protect the trees. Firewood collection for sales in towns and cities is an income generating activity usually carried out by landless youth. Degraded lands could be given to these young people to plant *Acacia tumida*, manage them and earn a living from them.

Mamby Fofana by one of his selected *Acacia senegal* trees.

It is possible to put an end to the destruction of **indigenous** tree species by planting **exotic** tree species on degraded lands. Sounds like a paradox, doesn't it?

Acacia senegal (gum Arabic)

When laymen talk of tree plantations in the Sahel region, they think of *Acacia senegal* for gum arabic production.

This is because they do not know about, or are reluctant to try other options. Both the World Bank and USAID are promoting plantations of these trees in Sahelian countries. Apparently, they are not aware of the following economic facts:

First, the market size for gum arabic is relatively small. It is estimated that exports from all gum producing countries in dry Africa (notably Sudan and Chad) are only around 60,000 tons per year. If production increases above current levels, prices will fall drastically due to the limited market whereas if prices are increased gum Arabic will be replaced by other cheaper gums.

Another concern is that the current average gum arabic yield in most Sahelian countries is 125 grams per tree per year. This translates to 78 kg/ha. The best price that a farmer can get for his gum arabic is $1.0 per kg, which would earn him $78 per hectare. Instead, if the same farmer can grow millet in his field, he will receive a yield of 300 kg/ha and a gross revenue of about $90 in regions of 400 mm rainfall per year— higher than the revenue from gum arabic.

The cost of digging demilunes, planting and plant caring is about $250 per hectare. A farmer must then wait five years to get the first gum yield, further delaying most economic benefits. Only then will he start gaining a gross revenue of $78/ha per year from the gum...

The high cost of gum arabic production coupled with the low income per hectare leads to planting *Acacia senegal* in degraded lands that are not cultivated, rather than in more fertile soils used for crops production. The best way to make gum arabic production more profitable is to bring about a very significant increase in gum yield. And this is possible.

In this pursuit, Mamby Fofana, followed by Albert Nikiema of ICRISAT attempted to select high yielding *A. senegal* trees and discovered that there were individual trees that originated in Sudan producing 800 g of gum per tree. Based on work done at the Central Arid Zone Research Institute (CAZRI), Albert Nikiema demonstrated that gum yield could be further increased through application of commercial ethylene products. The selected trees with high gum yields were grafted onto local *A. senegal* rootstock because the local germplasm is more tolerant to water submersion and droughts than the Sudanese germplasm.

If plantations with 0.8 kg of gum per tree (instead of the average 0.125 kg) can be planted, gross revenue of $500 per hectare—a very good income, can result. There are persistent rumors that in the Dourbali region of Chad, there are *A. senegal* trees that give about 4 kg of gum per year. If this can be confirmed, and varieties can be grafted on local germplasm (at the nursery or in the field as we did with Pomme du Sahel), then *A. senegal* plantations could become a rewarding enterprise at current farm gate prices of $1.0 per kg.

In the Sudan, there is a tradition of planting *A. Senegal* trees as a fallow crop to improve soil fertility. However, the plantation areas in the Sudan are diminishing due to economic and socio-economic constraints as well as demand for crop lands. In most of Sahel, gum arabic is collected from wild trees by nomadic tribes (mainly by women because men will not bother with the low income generated from the trees) in order to increase household income.

At present, I do not recommend planting *A. senegal* in Sahelian countries using local germplasm. There is however potential for this plantation crop if high yielding varieties grafted on local trees and technologies for increasing gum yield can be implemented. It is my hope that institutes like ICRAF (the World Agroforestry Center) will take this challenge upon them.

Tree domestication

Plant (and animal) domestication is probably the most ancient scientific activity of humankind. This action allowed the transition of the human race from the status of hunters and gatherers to that of settled, food-secured farmers and eventually to the creation of "civilization."

Tree domestication reached a peak between the 17th to the 19th centuries, following the expansion of colonialism and the discovery of new useful plant species such as cassava, cocoa, coffee, cashews, pecans, papayas, bananas, tobacco, coconuts, oil palm, rubber trees, and many more. This relatively recent wave of tree transfer and domestication has had a very significant impact on the economy of the tropical regions of the world.

Prof. Yossi Mizrahi, the tree domestication specialist at the Institute for Agriculture and Applied Biology of the Ben-Gurion University together with a Finnish NGO, discovered a novel method for rapid domestication of *Sclerocarya birrea* subspecies *caffra* (marula). First, rootstocks of marula were prepared in a nursery. Then a competition was declared among rural school children. The child who could identify the marula trees with the highest yields and the largest and tastiest fruit would receive a prize.

Suddenly, the researchers had hundreds of "germplasm prospectors." The best trees were grafted on the ready rootstocks at the nursery and taken to research stations for scientific evaluation. This approach yielded five new quality varieties of marula.

Unfortunately, colonialism concentrated on domestication of trees for the more favorable tropical climates and totally neglected the dry regions of the world. I am convinced that the domestication of fruit trees such marula, *Saba senegalensis* karité (shea nut), and others for the *dry regions* of the world would have an impact on the sustainable productivity of these regions equal to the impact of the genetic improvement of the rubber tree and the cacao tree in the economies of tropical countries. The case of Pomme du Sahel in India and in Niger is a prime example for the effect of trees domestication in dry regions.

For twenty-five years, I was the director of an institute that excelled in domestication of trees for arid lands. This institute (IAAB) was responsible for the domestication of many perennial species including jojoba (*Simmondsia chinensis*), the prickly pear (*Opuntia ficus indica*), various pitaya species (*Acanthocereus, Echinocereus, Hylocereus, Selenicereus* and *Stenocereus*), marula (*Slerocarya birrea*), various dryland eucalyptus species, ornamentals and halophytes.

Tree domestication is justified only if there is a market for their products. Often one needs to develop the market before or in concert with the domestication process. For example, it does not make sense to increase the yield of *Acacia senegal* through grafting of improved varieties if there is no market to absorb the extra production.

3. Rainfed crops

Thus far, I have described tree-based systems for dry Africa, geared towards combating desertification while, at the same time, creating income to farmers. In most of these systems, trees are intercropped with annual crops.

At present, about 98% of agricultural lands in the semi-arid tropical countries of Africa (with some exceptions) are used for the production of rainfed crops. Some of these crops such as groundnuts and cowpeas and to a lesser extent corn and sorghum are used also for cash. I will not describe grain cereals and groundnuts in any depth because these crops are being covered by ICRISAT, the CGIAR and the NARs. In any case, my job was to diversify traditional

ICRISAT mandated crops. Nevertheless, I will mention in brief some special varieties of millet and sorghum but the emphasis will be on dolich, dual purpose cowpeas, forage millet and roselle.

Dolich (*Lablab purpureus*)— the ultimate "Food Security" crop

Dolich growing from water stored in the soil of a receding lake

Dolich is a crawling legume resembling cowpeas. It originated in Africa, but it is mostly produced and appreciated on other continents such as South Asia, Australia and South America. The advantage of this species for the semi -arid regions is

in its deep tap root system that allows it to extract water
and nutrients from deep soil layers.

Selling dolich forage in the city in March—a time of forage
deficiency

Dolich can produce high grain and forage yields just from
residual moisture in the soil. In India, it is planted in rice
paddies in the dry season, thereby, utilizing the residual
moisture left after the rice. In Brazil and Nigeria, it is planted
together with maize spreading on residual soil moisture
after the maize is harvested. In Tahoua, Niger, it is planted
in sorghum fields growing on heavy slightly flooded soils,
in the soil of receding seasonal lakes and in dry river beds.
If intercropped with maize or sorghum, dolich should be

planted two to four weeks after these cereals are planted so it will not reduce grain yield.

Dolich grain yield ranges from 1–3 tons/ha, double the yield of cowpeas in the Sudano Sahel. Its forage yield ranges from 2–4 tons/ha, and its leaves are used as a tasty cooked legume. Another benefit is that dolich fixes atmospheric nitrogen at a rate of 15–40 kg per 1,000 kg of dry matter produced. Dolich thus provides protein, vitamins and minerals for human consumption, forage for ruminants at a time of high forage demand, and it improves soil fertility.

The good thing about dolich is that usually there is no need to apply fertilizers or to spray against pests and diseases (Pasternak, 2013).

The reason I refer to dolich as the ultimate food security crop is that when we plant it on water-saturated soils *after* the water of flooded areas recedes or *after* rice, we *are* **sure** *to get a good yield*. This is in contrast to conventional rainfed crops such as cowpeas planted at the beginning of the rainy season where there is no guarantee of a reasonable yield due to possible long dry spells and drought.

With the exception of Nigeria and Kenya, there has been no research on dolich in dry Africa countries. Moreover, very few if any development agencies have bothered to disseminate information about this exceptional crop. A case in point is the recently compiled list of legumes to be promoted in dry Africa published by AGRA. Not surprisingly, dolich was missing from the list.

Dual purpose (grain/forage) cowpeas

I have dedicated more research time to cowpeas than to any other rainfed crop. The reason is that this is one of the few high value cash crops produced in the Sahel. The forage is sold in large quantities to towns and cities, where residents traditionally raise livestock in their backyards and the grain has high value. Most improved cowpea varieties used in West African drylands have been bred over the years by IITA (the International Institute for Tropical Agriculture) in a Sudano region receiving more than 1,000 mm of rainfall/year. Breeding and selection for the Sahel region was done by local NARs using mostly IITA germplasm.

Cowpeas originated in dry Africa and it is likely for this reason that it is attacked by so many pests there—so much so that if farmers do not spray it, yield can be as low as 10% of potential. And farmers usually do not spray cowpeas. Thus, my selection was towards both drought and pest tolerant varieties. I started the selection process with about 60 varieties of cowpeas including some from Mexico. The selection process ended with two dual purpose varieties: ISV 128 and ISV 20. These varieties were actually landraces (varieties developed by local farmers) from the dry region of Zinder, the ancient capital of Niger, which receives about 400 mm/y rain. Both landraces that were improved at ICRISAT outperformed the IITA varieties in the drier regions of Niger in terms of grain and forage yields.

Millet

Grain millet

In dry West Africa, millet breeding poses a great challenge to scientists. Dr. Oumar Niangado, a veteran millet breeder from Mali, explains that each region of the Sudano Sahel has developed its own landraces that were selected over hundreds of years to fit the particular microclimate and soils of the region. This is why it is very hard to replace "local" varieties with "improved" varieties. But there is an exception called HKP millet. HKP millet is a *short duration* variety with high tolerance to head miner, a major pest. In regions with annual rainfall below 500mm it performed better than the local varieties.

Most rainfed varieties growing in the Sudano Sahel will start flowering only after the days become shorter. "Short duration" millet varieties (70–80 days from sowing to harvest) will flower (and consequently produce grain) before the "long duration" varieties. This is a significant advantage in regions with a short rainy season.

Forage millet

Forage is a major limiting factor for animal production in the Sudano Sahel. Pearl millet (*Pennisetum glaucum*) is a C_4 species with a very efficient photosynthetic system. In many countries of the world, millet is a forage crop but *not* in the Sahel where it originated, because there its food-insecure farmers place priority on grain production over forage production.

For two years, I evaluated five millet varieties for forage production. All tested varieties started flowering late (long duration), allowing more production of green biomass before flowering as compared with the common short duration varieties. The variety Malgorou gave a dry matter yield of 8.5 tons/ha when harvested at the dough stage in a year with 450 mm of rain.

Because we wanted to demonstrate this potential to farmers, I asked one of my workers to cut the millet in front of a group of selected farmers. This act almost ended in disaster. The farmers practically attacked the poor worker for harvesting the plant before grains were formed because for them, this was heresy. They consider millet to be an almost sacred crop. You are permitted to use the forage only after the grains are harvested even though at this stage, the forage is of very low quality.

While traditional farmers in Niger may resist production of millet for forage, commercial farmers everywhere in the Sudano Sahel dealing with animal fattening will welcome the idea. The question, of course, is who should share this concept and disseminate the seeds of this exceptional forage variety known as Malgorou. By the way, this variety can also be used as a dual purpose grain/forage millet because the weight of the straw left after the grain harvest is very high (6.6 tons/ha in a dry year).

Dual purpose (grain/forage) sorghum

Since forage is a very important commodity in the Sudan Sahel, it makes sense whenever possible, to shift the emphasis towards dual purpose varieties of sorghum. The sorghum variety from Mali called Malisor produces quality straw and at the same time gives a high grain yield in regions with low rainfall.

Roselle (*Hibiscus sabdariffa*)

Roselle is an annual hibiscus that like cotton and okra belongs to the *Malvaceae* family. Roselle's specific outstanding feature is its tolerance to very low fertility in the soil. It is also tolerant to drought and to low and high pH soils. For these reasons, it grows extremely well (relatively) in the poor soils of the Sahel. The main product of roselle is the succulent flower sepals that contain a high percentage of anthocyanin, the pigment that gives a red color to so many fruits and other agricultural products and is an excellent antioxidant. The dry sour sepals are used for the production of a popular healthy juice called *bissap* in West Africa and *karkade* in the Sudan and in Egypt. Like many rainfed crops of the Sudano Sahel, roselle is day-length sensitive—it is the shortening of days that induces flower formation.

After fermentation, roselle seeds are used in most of the Sahel as a tasty sauce. But the seeds can be also used for oil extraction and as concentrated animal feed. The tender leaves of this plant are widely used as vegetables. The flowers of varieties with green sepals are also used as vegetables. Roselle stems can be used for rope and paper production and the dry leaves for animal forage.

There is a good export market for dry roselle sepals in Europe and in the US where they are processed into herbal teas, various drinks, food coloring and cosmetics among other uses. Annual import of dry roselle calyces to these two regions is in the vicinity of 60,000 tons. Coastal African countries such as Senegal export about 1,000 tons per year. Exportation from landlocked countries such as Burkina Faso, Niger and Mali is problematic because of the need for overland transportation to faraway ports.

Roselle calyces ready for harvest

A serious concern is that roselle depletes the soil of nutrients. For this reason, farmers (mostly women) plant it at the periphery of cropped fields. But my experience has shown that if it is planted in zaï holes enriched with manure, one can grow it without fear of nutrient depletion. Most of the roselle produced is consumed in its country of origin by the farming family or sold in local markets. More research and promotion are needed to upscale the production of this crop as it is most suited to the soils and the climatic conditions in the Sudano Sahel.

Roselle as a leafy vegetable

In 2002 and 2003, I conducted a series of experiments to study the performance of roselle as a field crop. Three varieties were tested: one with purple sepals originating in Senegal, one with pink sepals from Niamey, Niger and a third with short pink sepals from Tanout—a dry region in East Niger. Dry calyces yield was similar among the three varieties, ranging from 450–500 kg/ha. However, the Tanout variety matured three weeks earlier and gave higher grain yields than the other two varieties.

In 2008, we discovered a single roselle plant that was not day-length sensitive. It flowered much earlier than the rest of the population. My guess is that it was a spontaneous cross between the Niamey and the Tanout varieties. This single plant became the origin of a new high yielding and high quality short duration variety that we named the "Sadore" variety. Sadore appears to be doing very well in the drier regions of Niger.

Roselle is not a new crop in the Sahel, but until now, this region has been unable to establish a solid export industry of this crop. Good quality roselle calyces can be sold for $1.0/ kg at farm gate prices, meaning about $450 gross revenue per hectare. Therefore, governments and development agencies particularly in countries bordering the sea should take the challenge of expanding exports of this crop.

Chapter 5

Irrigation is the solution

About 98% of Sudano Sahelian croplands depend on rain for water; only 1–2% of fields are irrigated. Crop failures due to droughts occur in two out of every five years. For example the Sahel bordering the Sahara, suffered from droughts in 2005, in 2009 and again in 2011.

In the Sudano Sahel, with its reasonable interstate road system, it is relatively easy to move food from high rainfall areas places such as Benin, Ghana, and South Mali to the drought affected areas to the north in times of need. But naturally, the price of this imported food is more expensive than locally produced food.

In dry Africa, *people go hungry after droughts not because there is no food. People go hungry because they do not have the resources to purchase the higher priced food.*

In order to buy food, desperate farmers will sell their possessions including farm animals and agricultural tools, eat the seeds that they have kept for sowing in the following season, cut and sell firewood and migrate to places where they can find temporary work. Thus, when the rains return, these farmers are left without the seeds or tools necessary to work the land or animals that provide cash. Therefore, it takes many years for a farmer to recover from a drought. And then when he does, another drought is imminent.

With such frequent occurrences of droughts, the rainfed agricultural system of the Sudano Sahel is not sustainable.

One can think of a number of strategies that can mitigate the effects of droughts. One of these is planting short duration crop varieties. For instance, there are millet varieties that mature in seventy rather than ninety days from sowing and can therefore produce grains even if the rains stop earlier than usual. Another strategy is reliance on trees and tree products. As already discussed, trees are much less affected by dry spells between rains than annuals and therefore have a better chance to give a yield under drought conditions. Yet another worthwhile strategy is water harvesting as in the BDL.

But irrigation is by far the most promising means for sustainable and profitable agriculture in the Sudano Sahel and in other dry regions of Africa.

In 2009, there was a severe drought in East Niger which caused millions to go hungry. One day, I decided to call on a farmer friend in a village on the road from Zinder to Diffa, a region where the drought was most severe. The village is located near a site with a shallow water table that is used for vegetable production. I asked him how he and his village were coping with hunger: "Hunger?" he said, "oh yes, I heard about it on the radio. But as you know, farmers in my village live from *irrigated* vegetables and cassava, and here, there is no hunger."

While the Sudano Sahel is a dry region, and people always associate dry regions with water scarcity, *this region does in fact have adequate water resources.* In addition to the 80 billion m^3/year discharge of the three major Sahelian rivers (the Chari, Niger and Senegal Rivers), there are billions of cubic meters of water in shallow aquifers and billions in artificial and natural lakes. In addition, the Nubian Sandstone aquifer that underlies the Sahel, stores about 65 million cubic *kilometers* of fossil water. This water is there waiting for the day when other water sources become overexploited.

Drip irrigation was developed in Israel. In the years that followed its invention, drip companies added many auxiliary parts to the drip system (fertilizer pumps, self-cleaning filters, irrigation controllers) turning drip irrigation into quite an expensive and complex system to operate. At that time,

I agreed with most experts that drip irrigation was not a suitable option for poor inexperienced farmers.

In 1982, I completely reversed my position about this when I toured the coastal area of western Egypt together with Egyptian and Israeli scientists. In the vicinity of Marsa Matrouch, we came to a small experimental agriculture station where the manager, who looked like a typical Egyptian *felaheen* (peasant), demonstrated a low-pressure drip irrigation system that was irrigating a small olive orchard. After hearing about drip irrigation, this man had bought 16-mm polyethylene pipes and small spaghetti-like tubes in Alexandria. Then he cut the spaghetti tubes into 20-cm pieces and immersed them in the 16-mm laterals, turning the tubes into drip emitters. He filled a reservoir constructed of concrete blocks and positioned four meters above the field with water pumped from a sand dune aquifer, thus devising one of the first low-pressure drip systems in the world. Needless to say, then and there I lost all my misconceptions about the exclusivity of drip systems. It took me sixteen more years until I was given a chance to develop low-pressure drip systems for the poor.

Now that we are aware of the huge potential of water that is available for exploitation, two questions come to mind:

1. If there is so much water, why is no one taking steps to utilize it?

2. Since water exploitation for irrigation is quite costly, which are the crops that can "pay" for the water, and what are the best irrigation systems?

The answer to the first question is that governments and others have indeed given much thought to irrigation (perhaps not in the general context described above), but so far, the performance of what has been done has not been satisfactory, discouraging further attempts for water exploitation. For example, Burkina Faso is a country of meager shallow groundwater aquifers due to the non-porous granite rocks that underlie much of its surface. Over the last thirty years, about 1,700 dams have been constructed to harvest the significant amount of runoff water. While the amount of water stored each year in these reservoirs amounts to 2.7 billion m³, *only 12% of the stored water is used for irrigation.* The vast majority is lost to evaporation and deep percolation (DGIRH, 2001). The government of Burkina Faso has yet to find an effective way to utilize most of the water stored in its artificial lakes.

The high cost of water development (and supply) in public irrigation schemes is yet another constraint for proliferation of irrigation. For example, the World Bank (2008) calculated that the cost of the water supply through public schemes for surface irrigation in Niger using Niger River water amounted to at least $20,000 per hectare. To complicate matters, many of the farmers that are using water from public schemes do not own the land that they irrigate. Therefore, they lack the motivation to "invest" in that land. For example, when I was trying to introduce low pressure drip irrigation to northern Ghana, I worked with three groups of farmers. Two of these groups owned the land and controlled the water. These groups readily accepted the new technology. However, farmers in the third group were "vassals" of the water authority of the Volta River, a government irrigation company. The introduction of the AMG technology to this group was a complete failure.

The failure of public irrigation schemes has led many to believe that small scale private irrigation is a more reasonable solution for water utilization. For this reason, governments, development banks and donors are now transferring their support from public schemes to private irrigators. Although there is increasing awareness among governments, development banks, the FAO and some bilateral donors of the importance of irrigation, the pace of its dissemination is too slow for my taste.

In regard to the second question (which crops to grow?), the answer is simple: Concentrate on crops that bring a high return per unit of water applied. The higher the cost of the water, the higher the return from the irrigated crop should be. As a rule, horticultural crops give the highest return for water applied. This is why most of my "crop diversification" activity at ICRISAT revolved around such crops.

Many people consider the "Green Revolution" that resulted from breeding superior grain varieties coupled with increased application of fertilizers as praiseworthy. So do I. But few are aware that in Asia, the birthplace of this revolution, success came as a result of *irrigating* the "miracle" varieties. Today 40% of Asia's croplands are irrigated, growing 70% of the staple crops, most based on Green Revolution varieties.

My agricultural research career began just before the onset of the Green Revolution when high yielding wheat varieties from CYMMIT were introduced to Israel. At that time, irrigating common wheat varieties was actually resulting in lower yields. Old wheat varieties invested the extra photosynthates (the products of photosynthesis) in stem and leaf production rather than in grain. This resulted in severe crop lodging (falling over) with a corresponding reduction of harvested yield.

The short-stem new wheat varieties did not lodge under irrigation and were able to invest the extra production that came from irrigation into the grains. In other words, the Green Revolution varieties turned wheat from a staple *rainfed* crop into an *irrigated* cash crop. *Today 80% of the green revolution high yielding wheat produced in India is irrigated* (Punjab National Bank, 2010). What I fail to understand is why the promoters of the new Green Revolution for Africa still insist in producing rainfed crops.

Technologies for small-scale irrigation

In the Sudano Sahel, about 80% of gardens are hand irrigated using watering cans, calabashes or buckets. The rest are surface irrigated.

In 1997 and again in 1998, I visited four Sahelian countries with the objective of finding ways to improve their agricultural performance.

I realized then that irrigated market gardens could benefit significantly from improved irrigation technologies.

I returned to Israel, and over a period of three years, we developed and tested a low-pressure drip irrigation system that was called the "African Market Garden" (AMG) at my institute in Beer Sheva. The AMG can be described in an analogy to the computer. It has "hardware" (composed of wells and boreholes, as well as a reservoir and pipes) and "software" (the garden management system) components. The management package includes land preparation, manure and fertilizer application, irrigation and fertigation, vegetable varieties and vegetable nurseries. As in a computer, the hardware is totally useless without the software.

At the very same time that we were developing the AMG, other organizations were also developing low-pressure drip systems, but their emphasis was on the hardware. Little attention was paid to the management package—the software. This resulted in many failures (Belder et al, 2007). Likewise, their attempts to produce a low-cost "affordable" drip system (Polak et al, 1997) were also unsuccessful because they did not take into account that *low cost* is often synonymous with *low quality* leading to bad hydraulic performance coupled with much maintenance. Bear in mind that maintenance is a serious problem in places where spare parts are not readily available and where a maintenance tradition is lacking.

Poor farmers simply cannot afford this type of "affordable" low cost system.

It took us ten years of trial and error in six Sahelian countries until we came up with the optimal configuration of the AMG (Woltering et al, 2011a, 2011b). We developed the following four models to fit a range of situations:

The "Thrifty Model" is based on a reservoir made out of used 200-liter oil drums elevated one meter above the ground. The system can irrigate an area of 80 m^2 if barrels are filled twice daily with water. Unfortunately, however, we found out that the effort of filling up the barrels by hand twice a day ,is greater than the effort invested in irrigation with watering cans. Therefore, this model has been discarded in most places except in situations where land and water are very limited such as in the Cape Verde islands. There, an NGO called Caritas is promoting the thrifty system. Currently, more than 2,000 units are operating, and that number is growing

The "Commercial System" is a small field 500–2,000 m^2 in size that receives water from an elevated concrete reservoir. The volume of this reservoir is based on the quantity of water that is provided to the field on a daily basis. In Niger, with its very high evapotranspiration rate (due to the hot and dry climate), the daily water requirement in the hot dry season reaches 8 mm, and the volume of a concrete reservoir that supplies water to a 500 m^2 plot is thus 4 m^3. After the reservoir fills up, the producer opens the tap and allows the reservoir to empty out, this being the daily amount of water required by the plot. This model is most suitable for learned small commercial farmers.

The "Cluster Model" is where numerous 500–1,000 m^2 plots are delineated in a larger field. Although water is supplied from a single source to all units, each producer farms his plot individually.

The "Communal System" is the most exciting model that we have created. In this system, the field resembles a pressurized drip-irrigated field. It is divided into plots given to individual producers varying in size from 100 to 1,000 m^2, depending on requests by the community or the donor. Each farmer prepares his/her plot giving the recommended doses of manure and complete fertilizer. Irrigation and fertigation (the practice of mixing a dose of soluble fertilizer in the water with each irrigation) is done by the community. The community is also responsible for treating the crops against insects and diseases, for purchasing farm inputs and for marketing. This system is ideal for women because a woman with a sick child or one who has just given birth need not worry about tending her field on a daily basis because her plot will be irrigated by the community.

Overall field size for the communal and cluster models ranges from 0.5 to 5 hectares, divided into plots according to the number of members. Due to the "economy of scale," investment per m² is only $1.50 in these two systems and due to the efficiency of both; the payback period is only six months.

The AMG is a simple easy-to-operate irrigation system. Still, to be sustainable, good quality equipment must be installed, producers must be well organized, spare parts should be readily available and producers must master the management package.

We have found that a minimum of three years of intensive follow up is essential to allow the proper "digestion" of the system by the producers.

AMG for small farmers

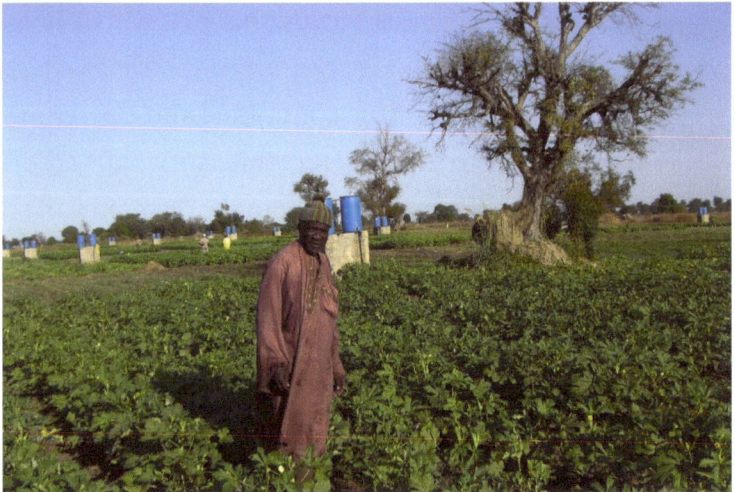

Communal AMG. Every farmer has a 1,000 m^2 plot

Water pumping and delivery

Private irrigation many times means that the individual farmer needs to seek out and provide his own water for irrigation. He needs to dig the well or drill the borehole, and then he needs to lift the water and deliver it to the field. Water supply is the trickiest part of a private irrigation system and it is fraught with problems. It can be very costly in labor if that labor is used for hand lifting and water delivery. If small motor pumps are used, the pumping depth is limited to less than ten meters, and the cost of fuel is very high. Additionally, motor pumps often break down and require much maintenance. If treadle pumps are used, pumping depth is limited to seven meters, the cost of labor is high, and these pumps too are often subject to breakdown.

The collaboration between Israel and the Vatican in the promotion of African Market Gardens in seven Sahelian countries was a most rewarding and interesting undertaking. ICRISAT worked under contract with MASHAV the Israeli international cooperation body in direct contact with the John Paul II Foundation for the Sahel of the Vatican. In a period of two years we installed African Market Gardens (AMG) in seven Sahelian countries. The technology picked up in Cape Verde in Mali and in Senegal and is spreading fast in these countries particularly through the Catholic NGO called Caritas.

This was a rare example on how people of three   religions can work together to benefit the poor of Africa. The director of the program was a Jewish Israeli (me), the technician who spread the technology was a Moslem and the recipients of the technology were Catholics....

The AMG irrigation system needs very low pressure for operation, which means that the system can use low energy water sources. Here are some examples:

Solar energy

Theoretically, solar operated pumps are the best technical solution for water pumping and delivery for poor small farmers in places (most) where there is no connection to electricity supply. Once installed, the energy is "free" and no less important, maintenance is almost non-existent. Electric pumps have an amortization period of ten years with very little breakage, and companies producing solar panels provide a twenty-five guarantee. The main problem with this technology is the initial high cost of investment. The good news is that prices of solar panels are decreasing. Indeed, prices fell by half between 2009 and 2011. Traditional DC Solar pumps operate best if pumping is done from shallow boreholes down to twenty meters deep and when low pressure is needed for irrigation, as in surface irrigation and the AMG.

The solar panels generate electricity as Direct Current (DC). Some companies have developed "solar pumps" that operate on Direct Current over a wide voltage range allowing for the constantly changing output of the solar panels due to cloudiness and hour of the day. These pumps are expensive, producing relatively little water pressure.

When drip irrigation is used one therefore needs to elevate the pumped water to reservoirs and irrigate the fields at low pressure from these reservoirs that provide a constant pressure resulting in steady water discharge. The reservoirs are expensive. An alternative is to store electricity in batteries. However batteries are expensive. They are stolen and have a relatively short life span.

Ran Zechori is a friend and owner of the African Irrigation Projects (AIP) company. He has been following closely the evolution of the African Market Garden and of solar operated pumps. He came up with a low cost inverter from

Direct Current to Alternate Current (AC) that supplies electricity at a steady voltage to conventional non expensive electric pumps. Now we can use "normal" AC pumps to pump water from various depths to irrigate fields (Including drip irrigation) at high pressure.

In addition we overcame the need of expensive reservoirs through the use of automated irrigation technology.

Solar energy is now ready to revolutionize small-scale irrigation in Africa

The California system: PVC lines are buried in the ground at a twenty meter spacing. A water distribution raiser emerges every twenty meters, providing water to a 400 m² area. Big savings in fuel and labor, good water distribution and higher yield— strongly recommended

Hydraulic gradient

In 2006, we introduced the AMG to northern Ghana. Gardens were established downstream of two dams, the Golinga Dam near Tamale and the Bauku Dam near the border of Burkina Faso. In both places, the dam water was fairly clean, allowing it to be used in low pressure drip systems without clogging the drip emitters. The difference in height of three to five meters between the dams and the fields below provided enough pressure to operate the AMG. There are dams that can supply clean water under low pressure to operate thousands of AMG units in many countries of the Sudano Sahel. This source of "free energy" should be not being overlooked.

Artesian water

One day in 2007, I was invited to conduct a survey on the potential of artesian boreholes as a source of water for market gardens by JICA (the Japan International Cooperation Agency)-Niger. Prior to this time, I was unaware that a huge confined aquifer (an underground water body closed from above by impermeable layers) exists in Niger, extending to Mali and Nigeria. This type of aquifer, called an artesian aquifer, is naturally under pressure and once a drill penetrates the impermeable layer and reaches it, the water rises to the surface through its own energy. *In other words, there is no need for pumping.* Further investigation has revealed that in Niger, there are about 200 boreholes that have already been drilled into the artesian aquifer and most are underutilized. Currently, water comes up to the surface running freely twenty-four hours a day, creating a small lake in the vicinity of a borehole.

The first demonstration site for the use of artesian water for an AMG in Niger was in a village called Yelou in the south. This project had the support of USADF (the United States Africa Development Foundation) and was on a five hectare AMG field irrigated with artesian water. Currently, it is providing a livelihood for 100 small farmers, mostly women, each making a significant income from a 500 m² plot. As I am writing this book, USADF is supporting the establishment of three additional AMG sites utilizing artesian water. I believe that drilling of artesian boreholes will provide free water for market gardens in the most remote areas of Niger.

Use of existing infrastructure

In Senegal, almost every village or cluster of villages has an autonomous supply of drinking water. Donors (notably JICA) and governments usually construct a fifteen to twenty meter high water tower attached to a water pump. In most cases, there is a surplus of water in these towers with significant water pressure that can be diverted for irrigation of an AMG. The water users must pay the municipality for the water but are exempt from pump purchasing, operation and maintenance. And, no less important, the water in most places is clean and will not clog the drippers. It is now possible to install hundreds of AMGs attached to village water towers in Senegal or in other places with these towers.

The California system

The California system was introduced to Niger by Jon Nougle from an NGO called "Enterprise Works." It is now spreading in other Sahelian countries. This is basically a water distribution system used mostly for surface irrigation. It consists of PVC lines buried in the ground with an outlet every twenty meters. This inexpensive method of water distribution saves about two-thirds of the labor and fuel compared with traditional surface irrigation systems. It also saves water and produces higher yields due to more targeted water distribution. Small farmers irrigating small parcels of land (0.5–5.0 ha) praise the California System and are ready to invest in it.

The Gaya system

Gaya is a region in south Niger with an average annual rainfall of 800 mm. Over a period of twenty years, a local farmers' cooperative developed an ingenious and indigenous integrated production system of fruit trees, rice, and onions.

How did this project get started?

In the 1980s, French Cooperation decided to introduce fruit trees to the Gaya region. They organized a group of farmers into a cooperative, installed a big diesel pump on the river and planted a range of fruit trees, mostly citrus species in a sandy soil. However, fruit trees did not do well in the sandy soil, so farmers started planting them in the heavier soil traditionally used for rice production.

After some time, the huge pump broke down and was replaced by small motor pumps owned by individual farmers pumping water from shallow boreholes, which was cheaper and easier to execute then central pumping of river water. Since these farmers were originally rice producers, they naturally planted rice between the young fruit trees in the rainy season. Then the idea was conceived of planting onions and other vegetables in the dry season after the rice was harvested, and this is how the Gaya system was born. Since availability of good soil was limited, the cooperative decided to allocate a 0.25-hectare plot to each of its members.

The Gaya system is synergistic. The rice destroys noxious weeds and soil diseases and suppresses nematodes introduced by vegetables. At the same time, rice enjoys the high fertility that remains after vegetable production. And the fruit trees enjoy water and fertility from both the vegetables and the rice.

One day, together with Saidou, my assistant, I traveled to the Gaya region in search of *Citrus volcameriana* seeds. This is a wild citrus species resistant to some soil diseases and is therefore used in many nurseries as "rootstock" (the plant part upon which a desired variety is grafted) for cultivated citrus. When searching for the volcameriana seeds, we came across a sign stating that we were at the site of the "Integrated Rice–Fruit Tree–Vegetable System." Saidou had a good hearty laugh. "It is impossible to grow fruit trees together with rice in the same field," he said. We got out of our vehicle, and to our astonishment, we saw a beautiful citrus orchard growing inside a rice field. This was the beginning of our long romance with the Gaya farmers' cooperative.

Gaya system in dry season—vegetables plus fruit trees

Gaya system in rainy season—rice plus fruit trees

We have been able to demonstrate that the *annual profit* from a 0.25 ha system plot in Gaya amounts to $1,000, enough to provide a decent livelihood for a farmer's family. In the Sudano Sahel, the Gaya system and its variations could be applied to the thousands of hectares presently growing just irrigated rice. The irrigation infrastructure is already in place. All that is needed is for someone to start the process of diversification.

Irrigated crops

1. Vegetables

I conducted research on irrigated crops mostly with vegetables and fruit trees because, as I mentioned earlier, horticultural crops bring the highest revenue per unit of water applied and per production area. However, I am aware that there is much research to be done on irrigated cotton, maize, wheat, cassava, soybeans, forage crops, sesame, bananas and other crops leading to the creation of a broader basis for large-scale irrigated agriculture in the Sudano Sahel. *To the best of my knowledge in this part of Africa there is no research organization yet studying irrigation of the above crops.*

In 2002, when I started working on vegetable production for the Sahel, I immediately realized that something was wrong. At that time, there was no breeding and selection of vegetables specific to the region, and farmers were buying seeds of the varieties sold by local traders, who imported them from Europe. The seed traders were and still are considered the "experts" on the vegetable varieties, selling the farmers whatever they happen to import or what was left over from the previous season. Many times, the seeds are old and their germination rate is very low.

Before we started the vegetables research, ISRA, the national agriculture research institute of Senegal had been the only research organization that attempted development of new vegetable varieties of many species specific for the region, but they had stopped due to budget constraints.

Most of the varieties grown by small farmers in the Sahel are open pollinated because farmers refuse to pay for the more expensive hybrids. (A "hybrid" in contrast to "open pollinated" is the result of a cross between two specific parent plants. Seed companies keep the "identity" of the parent plants secret so that only they can produce the specific hybrid.)

> We owe a debt of gratitude to Itsik Nir, the founder and president of an organic seed company in Israel called Genesis. Itsik was able to collect an impressive number of open pollinated vegetable varieties before they totally disappeared due to the strategies of seed companies that now produce mostly hybrids. He willingly donated seeds of many of our currently recommended varieties such as the Ein Dor melons and the True Gold sweet corn.

Because breeding is a lengthy process requiring many resources and time, I decided to take the approach of selection and purification of existing open pollinated varieties to allow me to reach a range of vegetable varieties adapted to the local conditions in **a short period of time**.

I concentrated **only** on open pollinated varieties to allow local seed producers and farmers to produce the seeds of the selected varieties by themselves.

In the following section, I will describe the reasons, ways and means for selection and purification of some of these vegetable varieties. The story is fascinating, but more importantly, there are worthwhile lessons to be learned from the original approaches taken in the selection and multiplication of these varieties.

Improved Violet de Galmi onions

To start the selection of adapted varieties, I first identified the most important vegetable species and varieties in the region. I found out that onions are one of these. Africans use them in sauces with almost every meal. The Violet de Galmi onion variety was bred by a French research organization called CIRAD (Centre de Coopération Internationale en Recherche Agronomique pour le Development) in the 1960s. It is a tasty, day-length insensitive variety with a reasonable storage life and good yields. Violet de Galmi is by far the most popular onion variety in West Africa. I tried introducing some new varieties from abroad including hybrids but found them to be inferior to Violet de Galmi.

After breeding was completed, the Violet de Galmi variety had not been maintained to keep it "true to type." Hundreds of small local producers had started to produce seeds without any formal training in this art. The result was hundreds of lines each with its own special characteristics, many of them negative.

In 2004, I compared the performance of forty-five Violet de Galmi lines that I had collected from all over Niger from the best seed producers and determined the two best lines:

one that gives very high yields but has a short storage life and the other (Line 28) that has a lower (but still reasonable) yield but very long storage life. Seeds of line 28 are now produced by the Ainoma seed company of Niger.

Icrixina—solution for rainy season tomatoes

Tomatoes will not grow and reproduce in the hot season of April–June when air temperatures rise above 40°C. During the rainy season from June to October, the tropical conditions (particularly the warm nights) markedly reduce tomato fruit set. As a result, the prices of tomatoes in the Sudano Sahel are very high between the months of May–November.

With these limitations, the most natural thing to do was to find out if, from among the thousands of tomatoes varieties that exist in the world, we could identify a variety that sets fruit in the rainy season. We tested twenty-five varieties that were supposedly heat tolerant and found that the variety called Xina gave the highest yield during the rainy season. This variety had been bred by ISRA (Institute Sénégalais de Recherches Agricoles) scientists. However, (and again) due to lack of maintenance, the quality of the fruit was very poor.

After five cycles of purification, we received an attractive firm and uniform fruit—a miracle variety that not only will start giving fruit in August, the peak of the rainy season—

but will continue giving yields for seven months thereafter. This is what African producers like. We called this tomato Icrixina, combining ICRISAT and Xina. Today, there is no shortage of tomatoes in Niger markets from August onward even though the prices are still higher than in the dry cooler season.

However, there is still a need to solve the problem of lack of local tomatoes during the period of June–August. We demonstrated that this problem could be solved by growing tomatoes in the relatively cool Aïr Mountains at a 2,000 m elevation.

Non-bolting lettuce

European lettuce varieties start flowering (bolting) the moment the ambient temperature rises above a certain threshold. In the Sahel, where lettuce flowers all the time due to the constant heat, farmers pick the lettuce leaves from the flowering stem. These leaves are small, hard and bitter. The late Dr. Dov Globerson was able to produce lettuce varieties adapted to the heat of the Israeli summer. In 2005, Dr. Globerson started giving training at ICRISAT on seed production to West African seed producers. At the same time, I introduced one of his low bolting tolerant varieties called Noga.

Although this lettuce did very well in the heat of the Sahel, the market did not like it because they were used to a yellow-green lettuce and Noga was dark green. I consulted Dr. Dov Globerson about this. "This is not a problem," he answered. "I actually have produced a yellowish lettuce for the Israeli market, but Israelis rejected it because they are used to dark-green lettuce...." When I asked him what name I should give to his variety, he answered, "Call it Maya, the name of my granddaughter, but on the condition that you will never commercialize it. I am giving this variety to Africa's producers free of any propriety right."

"Well, Dov" I answered, "you just took the words out of my mouth. This is the kind of strategy that I am promoting." Maya is now the best quality lettuce available in the Sahel, and its popularity is helping it to spread quickly.

Heat tolerant sweet corn

In the Sahel, the months of July–September just before millet or sorghum is harvested are known as the "hunger period." This is the time when most of the grain that has been stored from the previous season is exhausted, but the new crop has not yet been harvested. So, people go hungry during these months.

Farmers plant grain corn in early June and harvest it in September, selling the cobs for roasting at the peak of the hunger season. It came to my mind that sweet corn, a food that is unknown in the region, could replace grain corn as a food during the hunger period. An old open pollinated

American variety called True Gold did much better in the rainy season of the Sahel than the hybrid varieties and other open pollinated varieties that I tried, because it was more heat tolerant.

Sweet corn is totally unknown in most of the Sudano Sahelian region, and people warned us that it will be difficult to introduce it. But we found this not to be the case. Everywhere that sweet corn was planted, it immediately became a success, and it is now spreading in the Sahel. This crop offers an added advantage in the green stems and leaves which are used as quality forage after the green cobs are harvested.

The Uberlandia carrot

Carrots are very popular in the Sudano Sahel. However, most varieties require low temperatures (that are not available in the region) to induce flowering. ECHO (Educational Concerns for Hunger Organization) is an organization based in Florida that provides seeds of vegetables and other crops to Christian organizations and individuals that operate in developing countries. It was in one of their pamphlets that I first learned about the "Uberlandia" carrot variety that had been developed in the state of Minas Gerais in Brazil. This carrot produces seeds in tropical climates a few months after planting *without the need for cooling.*

Uberlandia carrots producing seeds in the Sahel—a breakthrough

We tried the Uberlandia variety, *and we were successful. It is now possible to produce carrot seeds in the Sudano Sahel.* This is a breakthrough.

To induce flowering, the Brazilians recommend lifting up the carrots when they are four months old and replanting them. In the Sahel, they sow the carrot seeds in November and pick up the crop in March when temperatures begin to climb to the mid-forties. Lifting the carrot roots and replanting the carrots at this time of the year causes much damage to the plants. Fortunately, we found out that in the Sahel, there is no need to lift the roots and replant them to produce seeds. Just sowing a carrot nursery and planting the young seedlings at 15 x 20 cm spacing is enough to induce 100% flowering, resulting in a high seed yield.

Konni okra—early maturing, year round production, high yield of good quality

The Konni Okra

Okra is one of the most popular vegetables in Africa. Among the reasons for its popularity is that the main dish in
the Sahel is a thick porridge made out of flour from grains and starchy crops (millet, corn, sorghum or cassava). These thick porridges have little taste. Taste is added using a wide range of vegetables, fish, meat and spices.

Okra is a very important sauce ingredient. Okra can be grown under rainfed conditions, and women are usually its producers.

They cut the okra fruit into small slices and dry it in the sun. This dried okra can last for a year or longer.

When IPALAC was operating from Israel, we used to organize a workshop on trees for arid lands every year. In 1999, Lowell Fuglie and his charming wife, Caroline, were invited to give a presentation on moringa. On this occasion, Lowell told us about a moringa variety called PKM 1 (Periyakulam 1) with very high pod yields and with a fast growth rate that had been developed in the region of Tamil Nadu in India. Based on this information, I invited one of the developers of this "miracle" moringa variety to the IPALAC 2001 workshop to speak about it with a request to bring some of its seeds with him. These seeds served as the basis of the proliferation of PKM 1 in Niger and the Sudano Sahel. We discovered that PKM 1 produces three times more biomass than the "local" varieties. In addition, its leaves are much tastier, and it requires less than an hour of boiling to get rid of the tannins, whereas local varieties need to be boiled for up to three hours—a "miracle" variety of a "miracle" tree.

The rainy season in the Sahel lasts only 3–3.5 months. Early flowering varieties are better suited to the short rainy season than late maturing ones producing significantly higher yields. This fact is true for the short duration millets, and it is true for any other crop produced in the rainy season.

Our vegetable breeder from AVRDC, Dr. Sanjeet Kumar, decided to concentrate his efforts on breeding okra for the Sudano Sahel. He started by comparing under irrigation the performance of 150 accessions that he introduced mostly from African origin.

At that time, we were told that there is an exceptional okra variety that we "must try" in the area around the town of Birnie N'Konni in Niger. We bought seeds of this landrace in the local market and discovered an amazing variety that outperformed all the 150 accessions introduced by Dr. Kumar. *We called it the Konni variety.*

Konni is day length insensitive, meaning that unlike most day length sensitive okra varieties, it can flower and produce fruit all year. The first fruit is harvested fifty-two days after sowing. This is a dwarf plant that invests energy in generative rather than vegetative production, resulting in high yields. Thank you, Birnie N'Konni for this outstanding variety.

Moringa

Moringa oleifera is the last but not least of the vegetable species that I will describe. Moringa originates in tropical India (where they eat the green pods) and has spread all over the tropical world. In north Nigeria, Niger, and the Philippines, people consume the leaves, not the pods.

Moringa leaves are probably the most nutritious vegetable on earth. According to Mark Dafforn, my late friend from the American Academy of Science, moringa leaves have four times more Vitamin A than carrots, three times more iron than spinach, four times more calcium than milk, seven times more Vitamin C than oranges and as much protein as milk. Moreover, the leaves are very tasty if one knows how to prepare them well.

Moringa market in Niamey—the most popular vegetable in Niger

In Niger, moringa is the most popular vegetable. It outsells onions and tomatoes. Nigeriens are practically "addicted" to this food. In the Gambia and Senegal, moringa is called *nebeday*; a distortion of the English word "never dies." It is indeed difficult to kill a moringa tree. A tree can suffer long periods of stress, losing all its leaves and giving the impression of a dead plant. However, when stressful conditions are alleviated, moringa trees begin sprouting and growing. This is a very important trait for a perennial plant that is usually produced under sub-optimal conditions.

One-year-old moringa PKM 1 ready for harvest

The first time I heard about moringa was from the late Lowell Fuglie, the "apostle" of moringa. Lowell wrote a book called moringa the "Miracle Tree" (Fuglie, 1999) that is compulsory reading for all who are interested in this species. (I would add to this a book called "Moringa" written by Luis R. and Lalaine Villafuerte.)

In my opinion, moringa needs to be aggressively promoted in all of sub-Saharan Africa because it offers the best solution to vitamins and micro-elements deficiency called by many "the hidden hunger." It provides proteins and calories. It is also a good income generating crop. Many people in Niger eat moringa on a daily basis and the demand for this product is very high.

Moringa can be produced without or with irrigation. If it is not irrigated, it will become dormant during the dry season and yield during the rainy season. But if it is irrigated properly, one can have about ten harvests per year.

Aside from moringa, leafy vegetables are very popular in Africa. Many are still collected in the wild, but in very large quantities. Examples are leaves of baobab, various *Corchorus* species, and in Niger and the Sudan, leaves of the wild legume, called *Senna obtusifolia.* In Niger, one does not came across nutrient deficiency among children as compared with its neighbors. I suspect that the reason for it is the huge consumption of leafy vegetables, (cultivated and wild) by the population. Recently, many donors have started promoting "Nutrition Projects." They are educating women on the importance of a balanced diet, which is fine. But at the same time, these projects should be promoting massive production of moringa and other vegetables, as well as pulses. If they fail to offer highly nutritious food at the same time that they are preaching nutrition, their message will be blowing in the wind.

To augment this chapter on vegetable varieties, I have provided a table in Appendix 1 with the names and the major traits of the varieties that we have selected. Seeds of these varieties can be purchased from the Ainoma Seed Company in Niger (masalifou@yahoo.com).

To guarantee the availability of the selected varieties in all the corners of the Sudano Sahel, we have trained hundreds of farmers in the art of vegetable seed production. These small seed producers are now spreading our selected varieties and other varieties in Niger, Mali, and Burkina Faso and very soon in Senegal.

Post-harvest treatment of horticultural products

The term "post-harvest" encompasses all the treatments given to agricultural products after they are harvested. It includes storage, preservation (drying, canning), packaging, cooling, heat treatment, sterilization, controlled atmosphere, washing and artificial ripening. Except for the process of drying of products such as okra, cabbage, tomatoes, onions and moringa leaves, the post-harvest treatment of produce in the Sudano Sahel is a neglected field.

Onions and potatoes are two very important products sold nationally and exported regionally in this region. These two products are usually harvested in February–April, and because of the lack of appropriate storage facilities, these vegetables are sold immediately after harvest. This results in substantial price variations within the calendar year. For example, in Niger, onion prices can climb immediately after harvest in March are 8¢ per kg. Prices climb to 82¢ per kg in the month of October.

Both onions and potatoes can be stored for up to eight months after harvest. In the developed world, technologies exist for their efficient storage. These, however, are very expensive and consume a great deal of energy. For these reasons, they had not been introduced to the resource-poor farming communities of the Sudano Sahel.

Thus, we conducted research on onion storage. When farmers use traditional ineffective storage facilities for onions bulbs, spoilage after three months can be as high as 70%. We introduced an "old fashioned" technique of storing bulbs between layers of straw, combined with appropriate preparation of the onions for storing and reduced bulb spoilage to about 10%. Recently, we introduced a new low-cost onion storage technology that uses local materials, straw and passive solar ventilation (Pasternak et al, 2013) where onions can be stored for six to seven months without spoilage

Currently, there remains an urgent need to develop cost-effective storage for potatoes and to introduce a series of post-harvest treatments that will add value to horticultural products.

2. Fruit trees for the Sudano Sahel

The process of introducing fruit trees to any particular region is tedious and requires an extended period of time. Over the last fourty years, several French organizations have tried to introduce a range of fruit tree varieties to the Sudano Sahel, mostly through the local NARs. Fruit tree varieties were introduced to ISRA-Senegal, IER-Mali, INERA-Niger and IRAD in Cameroon. However, the number of species has been very limited— mainly mangoes and some citrus. Since fruit trees are quite rare in the Sudano Sahel, particularly in Niger from where I operated, we decided to introduce and test a wider range of fruit tree species and varieties for the Sahelian region.

In 2003, we established a drip-irrigated two-hectare mother plantation. Then I started traveling around the world (to Australia, India, Israel and West Africa) to collect and bring back the best commercial varieties with a likelihood to survive in the Sahel. We also established good relations with the US Department of Agriculture and were sent scions of sweet tamarind (USDA-Florida) and cuttings of figs (USDA-California). In addition, we introduced many varieties of mango and citrus from other collections in the Sudano Sahel. We built a spacious 2,000 m² regional nursery as well as a 1,000 m² special nursery for Pomme du Sahel and started multiplying the best selected varieties. In total, the Sadore nursery produced about 100,000 grafted fruit trees each year, and there was no problem selling them.

Not all of the introduced tree species have survived, partly due to the low pH and high aluminum content of the station soil and partly due to the very high temperatures of the Sahel. The names of the species and varieties in the ICRISAT-Niger fruit tree inventory are listed in Appendix 2. Plants and scions from this collection can be ordered from the ICRISAT (s.abdoussalam@icrisatne.ne) or from Ainoma (masalifou@yahoo.fr).

Producing and selling fruit tree plants is one of the most effective means of generating income on a national scale. Six years after planting, a fruit tree provides an annual income of \$20–\$40. The first 100,000 trees that were planted from the Sadore nursery have generated from year six onwards a minimal annual revenue of about \$3 million. And 100,000 trees were added each year just from ICRISAT's nursery, compounding this figure.

Over a period of ten years, we trained hundreds of nursery men and women in the art of producing and grafting various fruit tree varieties. In many instances, the nurseries built or upgraded after this training have become very profitable enterprises because the demand for quality grafted fruit tree plants far exceeds the supply.

Tree crops requiring irrigation in the Sudano Sahel

The most successful fruit tree varieties we have introduced are described in Chapter 4 under the sub-title "Rainfed tree crops." All fruit trees listed in this chapter can grow in the Sahel with or without irrigation. However, irrigation should significantly increase fruit yield and, in most cases, the profitability of the crop. In fact, many fruit trees that are suitable for the Sahelian climate and soils (such as various citrus species, bananas and papaya) will need to be irrigated to produce a reasonable yield.

In this section, I present the case of the Red Bangladesh papaya variety because of its role in income generation and nutrition, and I give a brief description of cassava and bananas due to their importance for food security and income generation.

Red Bangladesh—a papaya for the Sahel

Early investigations revealed that papaya was not doing very well in the hot Sahel even though it is heat and salt tolerant. It seems that the combination of high temperatures, high radiation and very low air humidity during the dry hot season is harmful to this species. In 2007, we imported seeds of six varieties of the "solo" type, the most popular commercial variety in dry West Africa to attempt to overcome this problem. In addition, we introduced two varieties from Brazil, one from Taiwan called "Red Lady" and one from tropical Bangladesh.

Red Bangladesh papaya. Heat tolerant, dwarf, very tasty, long shelf life

All varieties except for the one introduced from Bangladesh disappeared after one year. They just could not tolerate the harsh conditions of the Sahel. We multiplied the seeds from the Bangladesh variety and found out that the fruit of some of the "decedents" had a red color.

We assumed that these plants with the red fruit were derived from an accidental crossing between the Red Lady from Taiwan and Bangladesh varieties and called it "Red Bangladesh." This variety is a high yielder with a very tasty fruit that has a relatively long shelf life. After four additional generations of purification, we came up with fairly uniform semi-dwarf trees yielding more than 50 kg fruit per year.

Bananas (*Musa acuminata and Musa balbisiana*)

Bananas are heavy yielders. The yield of one hectare under irrigation is normally around 20 tons or more. This fruit is very popular in Uganda, where it provides 70% of the calories consumed by the rural population. In Mali, they are now planting large areas of irrigated bananas, mostly in their Sudano region. My main reason for not working on bananas in the past was that I was afraid that this crop would not tolerate the heat and dryness of the Sahel. But I was proven wrong. The region of Matam in Senegal is a typical Sahelian region and, there, irrigated bananas are growing over a large area using Senegal River water for irrigation.

The introduction and proliferation of irrigated banana plantations will be enhanced by planting high quality germplasm coming from tissue culture laboratories that already exist in West Africa.

Large irrigation schemes in the Sudano Sahel should look for crops such as corn, wheat, soybeans, cotton, cassava and bananas that occupy large irrigated areas. I must admit that over the last ten years, with the exception of cassava, I did not concentrate on irrigation of the above crops.

Cassava (*Manihot esculenta*) and other irrigated crops

Cassava is one of the three most important starchy crops in the world (the other two are wheat and corn). In tropical regions, it grows without irrigation producing a yield of about twenty tons of roots per hectare. Nigeria is now the biggest producer of cassava in the world. It is also the home of IITA (International Institute of Tropical Agriculture) that is conducting **excellent** work on cassava breeding and dissemination.

Cassava is tolerant to high temperatures and to poor soils. For these reasons, it can grow well in the Sahelian region— but only if irrigated. We compared yields of twenty-seven IITA cassava varieties under irrigation and found very large differences in yield between varieties. The best "sweet" variety, meaning a cooking variety, gave a root yield of fourty tons/ha. In the Sahel, this means gross revenue of $17,000 per hectare.

Cassava (front) and bananas under irrigation in the Sahelian region of north Senegal

This is food security combined with income generation.

With the expansion of irrigation in semi-arid Africa, cassava will become one of the major and most profitable irrigated crops. Furthermore, maize, rice, groundnuts, sesame, soybeans, sugar cane and vegetables for processing such as sweet corn and green beans, in addition to various forage crops will also be produced under irrigation. These crops together with fruit and vegetables will become the basis of a prosperous irrigated agriculture in the semi-arid regions of Africa.

Chapter 6

Realizing the Potential

In previous chapters, a varied assortment of technologies and crops have been described that could become the basis for agricultural transformation in semi-arid Africa. Most of these proposed crops and technologies are simple to implement and their costs are low. There are a host of solutions for agencies that do not want to invest heavily in the development of agriculture, in addition to ones for more costly intensification activities. The commonality between most crops and technologies described is that all have been successfully tested and their feasibility proven.

The first consideration for a planner of agricultural development is how to guarantee *sustainability*. Everything is built around this concern. Sustainability depends on many factors such as the **economic gain** from the specific development activity that will encourage investment by farmers for the up-scaling, the necessary skills of the farmers, the organization of the producers, the ready availability of inputs, market availability and connection to markets, environmental sustainability, effective extension services, means for the diffusion of know-how and good governance.

No project should be started before an in-depth analysis of the various aspects ensuring sustainability is carried out.

When planning a new development program, the first step should be to identify *all* the related factors such as availability and quality of natural resources (water, soil, climate, etc.), relative advantages, the social structure and level of education of the rural population. The integration of all these factors will result in a clear plan for development. The second step should be to design a series of activities that should be carried out, and through a process of elimination, to select the best options with which to proceed.

However, many planners do not take this approach. Most commonly, they identify the main crops and technologies that *are already being utilized in the region in question and develop a program to capitalize on these crops and the technologies.* For example, onions are a major export item in Niger. In 2001, the government, using World Bank funds, started a program for the promotion of agricultural exports from Niger, onions being the major item on their list. A few years later, USAID initiated a regional (including Niger) "Value Chain" program for onions, and in 2011, a new agricultural exports project financed by the World Bank was initiated, onions again being a priority crop. All this promotion of onions is because they "are there," and it is therefore "safe" to invest in this crop. While all this attention on onions is fine, there are many other export crops such as tomatoes, hot peppers, melons and watermelons that have ready markets in Nigeria, Benin and Ghana.

The conservative approach by the above planners is totally opposite to the one taken by Fundación Chile. Following a comprehensive analysis, as just described, innovations such as salmon and wine that were not pre-existing. were introduced by Fundación Chile.

The success of any enterprise or development program depends on five factors in order of importance:

1. **Vision**—the setting of clear goals and objectives

2. **Know-how**—the knowledge of technologies, capacity building and extension, marketing mechanisms, etc.

3. **Organization**—the work plan that is carried out at all levels: at the individual level, on a sub-regional basis and through public-private partnerships

4. **Human competency**—the staff which should be professional and dedicated

5. **Financing**—the necessary financial resources (listed last because if the first four elements are missing, no amount of money will be effective)

1. Vision

My list begins with vision because the vision of any organization or program (including that of governmental groups) eventually affects the outcome. From my experience, I recommend setting the vision as high as possible. For example, instead of the common vision of "poverty alleviation," strive instead for "poverty eradication." Instead of "food security," strive instead for "economic security." A single short-term program might not allow you to accomplish the far reaching vision that you have set up, but it will bring you a step closer to it. So please never lose sight of your grand-vision, and never compromise.

2. Know-how

In the last decade of the 20th century, IFAD (the International Fund for Agriculture Development) financed a relatively large project of stone lines or bund construction (technologies for water harvesting and for reducing soil erosion) in the Central Plateau of Burkina Faso. After the project started, a local scientist with the support of the University of Wageningen and IFAD conducted a series of experiments that determined the optimal distances between stone lines as a function of the slope of the area among other factors. Had this research been conducted **before** the project started, results would have been significantly better than those that were achieved.

Research and agricultural development

Coming from the world of research, it is only natural for me to elaborate on this issue and its importance for agricultural development. From my experience, research should always precede and lead development. All the "mistakes" should be made in the research phase—not during the development project. A positive example *of research preceding development* organization is CLUSA, who has recently started implementing a USAID "Food Security" program in Niger and in Senegal. Their agricultural activities are mostly based on research that was carried out at ICRISAT-Niger.

Not only should R&D precede development, it should be an *integrated* part of the development. Economic and socio-economic research in particular is extremely important at all stages. Economic research identifies the "weaknesses" of the technology under development, resulting many times in abandonment of a technology or in a radical change of that technology to fit reality.

An example:

When we began testing the African Market Garden, we developed two models. The "thrifty" model was based on a 200-liter barrel reservoir irrigating 80 m² of land, and the "commercial" model was based on a concrete reservoir irrigating 500 to 2,000 m² of land. The economic and socio-economic study showed that since the barrel in the thrifty system needed to be filled by hand (or by a treadle pump), there was no gain in labor savings compared with larger drip-irrigated gardens where the reservoir is filled by a motor or a solar pump. As a result of this study, we stopped promoting the thrifty system.

194 | Dov Pasternak

Whenever I have asked a bilateral donor to allot money for research to support a particular program that they were financing, their answer invariably has been: "Our country is already giving money to the CGIAR and the national research system and, therefore, we do not think that it should allot research funds to these organizations for the specific project in question," or "It is not the objective or the duty of my project to support or conduct research." Again, in my opinion, these answers are indicative of the lack of understanding of many developers of the role of research in agricultural development.

Research should and could be a source of innovation that so many times would completely change an existing situation. For example, the Central Negev Highlands in Israel was empty of agricultural settlements for many years because of lack of fresh water. Research initiated under my guidance in 1971 demonstrated that it was possible to use the relatively highly saline water that abounds in the region for irrigation. Based on this finding, Israel invested many millions of dollars in drilling new saline water wells and in building new agricultural settlements that use saline water for irrigation.

One can make a distinction between basic, applied and adaptive research:

a. Basic research is geared towards understanding basic questions and phenomena of nature and does not necessarily end up with immediate application of the findings.

b. Applied research is geared towards optimization and/ or evaluation of a production system or a component of this system, many times using principles defined by basic research.

Many scientists especially those in academia look down upon applied research. Unfortunately, this attitude often discourages organizations and individuals from conducting such research. But one must realize that agricultural research is geared towards serving an industry, and by nature, it is mostly applied research. Applied research can make its way into excellent scientific publications. For example, our work on the development of the African Market Garden ended up in three papers in distinguished publications including a publication in the highly regarded Proceedings of the National Academy of Science (Burney et al 2010, Woltering et al 2011a, Woltering at al 2011b).

c. Adaptive research is geared towards fine-tuning of existing technologies to a new environment, and it is usually carried out by extension workers in collaboration with scientists.

The CGIAR encourages its scientists to concentrate on basic research or what they call "upstream research." In my opinion, this is a wrong approach. The purpose of CGIAR is to reduce hunger and poverty in the developing world, and it is financed mostly by development agencies. Therefore, this organization should offer

196 | Dov Pasternak

concrete and fast solutions that pure basic research might not be able to provide, at least not in a short period of time. A blend between basic and applied research is probably the best avenue to take.

The biggest achievements of the CGIAR system are the superior varieties that have been bred by its various centers using *conventional techniques* of plant breeding. And conventional plant breeding is a classic example of a "downstream" (applied) research not of "upstream" (basic) research.

Research for development (R4D) versus research and development (R&D)

The term R4D has been recently introduced into international and national research jargon, replacing the "old fashioned" R&D. For me, a straightforward interpretation of R4D is "we scientists are here to do the basic research (for development), while 'others' should take our results and conduct development." I think that this approach, again, is a grave mistake. An applied research scientist (and agricultural research is mostly *applied* science) must be deeply involved in the first stages of the application (meaning development) of his research findings for the following reasons:

- The researcher is more interested than anyone else in the successful implementation of his/her findings and if allowed, it is the researcher who will become the driving force behind the up-scaling of these findings.

- He is much more fluent in his technology than those involved in the implementation and, therefore, he is able to best guide others in implementation.

- And most importantly, any research finding reaching the implementation phase needs to be continuously refined and modified to fit the realities of the field. The scientist who first developed the technique must be there in the initial stages of application to identify the weak points and conduct further research in order to overcome the hurdles encountered in the process of implementation.

This pilot R&D phase can take many years. For example, in Israel we worked on a small-scale irrigation technology from 1998–2001 based on low-pressure drip irrigation called the African Market Garden. It took a dedicated team of ICRISAT scientists and technicians **ten** additional years of R&D in five Sahelian countries to devise satisfactory systems that could be maintained by farmers' groups.

Unfortunately, often the attitude of those who advocate R4D matches the attitude of development organizations. Many people in these latter organizations perceive scientists as ivory tower dwellers and prefer them to keep their distance, leaving the "real" work to the "experienced" and "competent" developers.

R&D Centers

For many years, most agriculture research in Israel was carried out by scientists of the local NAR called the Volcani Center. Volcani had regional research stations all over the country. However, these stations were like closed fortresses that had little to do with the specifics of a region's agriculture. All the stations were fully controlled by a central facility, not far from Tel Aviv. In the mid-1970s, due to their increasing dissatisfaction with the Volcani Center performance, Israeli farmers rebelled and started creating their own "Regional Research and Development Centers." These centers are fully controlled by the farmers and deal with problems specific to each region. In addition to applied research, they are engaged in training and extension, in the establishment of pilot projects, and in many other regional activities with direct benefit to farmers. The R&D centers employ Volcani scientists and scientists from other institutes to conduct research in the centers.

These regional R&D centers are today the engines for the agricultural development of Israel.

It is only now, as I am writing these lines, that I realize that I have actually created a regional R&D Center for the Sahel at ICRISAT-Niger. This center has introduced and created new technologies and crops, trained farmers and technicians, provided new propagation material (seeds and plants) and developed pilot projects that tested the new technologies, varieties and approaches. To the best of my knowledge, there are no such R&D centers elsewhere in Africa. I have no doubt whatsoever that the creation of African R&D centers can become a most important tool for assisting agricultural development.

Agricultural education

Between 70-80% of the inhabitants of dry Africa make their living from agriculture. It is now commonly agreed that agriculture will be the driver of economic development in Africa. However, the transformation of agriculture can only be carried out by a generation of farmers well educated and skilled in a wide spectrum of agricultural activities. Practically, this means agricultural education at both the primary and the secondary school levels and continuing at the higher education level—not only for urban scientists and technicians but also for farmers. But the reality in Africa is that the number of agricultural schools in each particular country can be counted in one hand (in Niger there is only one), while schools for business administration abound everywhere.

Until now, most African universities that teach agriculture have concentrated their efforts on traditional rainfed crops, soil fertility, traditional animal husbandry and agroforestry. These universities (and the NARs of Africa) should introduce additional topics such as irrigation, horticulture, intensive livestock production, dairy farming and intensive poultry production.

In Brazil's northeast, scientists have found solutions for profitable development of their semi-arid tropics that are similar to those of Africa. Africa could benefit by sending graduate and post-graduate students to this place.

Capacity building

In 2003, we advertised for two technicians on irrigated horticulture in newspapers in Burkina Faso. About forty technicians applied - *none with qualifications on irrigation and/or horticulture.* I repeated the same exercise in Senegal in 2009. Again, we received about forty applications from qualified technicians, none of whom had experience on irrigated horticulture.

For years, the Sahelian higher education system has concentrated on teaching and training in a very narrow field of agricultural activities; mainly rainfed crops, animal husbandry, agroforestry and soil science, with almost nil emphasis on irrigation, horticulture, dairy and intensive poultry production. The lack of know-how of intensive agricultural production runs throughout the whole "chain of command" starting with scientists, going through technicians and ending with farmers.

I once separated the budget of a new African Market Garden project that I helped construct into two components:

The set-up and operation costs and the budget for capacity building. For each dollar invested in setting up the garden and operating it, we allocated two dollars to build the capacity of the producers and for follow up.

There are not enough professional technical people in Africa, and this is the Achilles heel of most agricultural development programs. While farmers may be reasonably trained in conventional rain-fed agriculture crops and systems, they have very little knowledge of other agricultural fields, particularly about irrigated agriculture. It should also be taken into account the fact that most farmers finished only primary schools and many are illiterate. These farmers will require much more training to digest a specific technology than literate and experienced farmers.

Building up the capacity of the rural population is the best guarantee for the sustainability of any agricultural development program.

The following is an example on the impact of knowledge transfer on development:

In one of our research programs at ICRISAT (Woltering et al 2011b), we noticed that proper crop management of the **traditional** market garden can more than double yields. In 2010, the Arziki program in Niger trained some 340 producers on improved vegetable production. (When asked these producers if they had received training in vegetable production in the past, their answer had been a definite no.)

These trainees, in turn, trained a total of 10,000 producers in their corresponding villages. An impact analysis of the outcomes of the training was carried out in June 2012 when about 60% of the producers were interviewed. About 70% of the trainees (7,000 producers) have subsequently practiced improved vegetable production Results are summarized in the table below:

2010 (baseline)			2011		
Area (ha)	Production (tons)	Yield (kg/ha)	Area (ha)	Production (tons)	Yield (kg/ha)
212	4132	19,511	359	14,850	41,370

Not only yields were doubled but the trained growers also increased the area of vegetables from 212 to 359 ha by using fields that already had the infrastructure for vegetable production (fences, boreholes, water reservoirs and pumps) but were not being used because of lack of profitability.

This is a very good example of what training farmers without any additional investments can do. Moreover, producers confirmed that their family consumption of vegetables grew by a large share in 2011 compared to 2010.

The Service Centers

The "Service Center" is a new approach for supporting large agricultural development project. It came out of the need to upscale the TIPA (another name for African Market Garden) project initiated by the Israeli Embassy in Senegal.

Over a period of five years, the Israeli Embassy in Senegal, together with a number of NGOs, introduced about ten TIPA clusters in Senegal. TIPA recently took a new turn when the Italian Cooperation, the Israeli Agency for International Development Cooperation (MASHAV) and the Ministry of Agriculture of Senegal signed an agreement to install 400 hectares of TIPA in Senegal at about 80 sites.

Carrying out this task requires a huge effort to build the capacity of technicians, village animators and producers in the art of drip irrigation and vegetable production in the provision of inputs and in marketing garden products.

Construction of "TIPA Service Centers" is enabling the project to move forward. The Service Centers performs the following tasks:

- Training of technicians, animators (see below) and producers in a wide range of topics related to drip-irrigated horticulture

- Carrying out follow-up activities and consultancies at all sites that received the initial training for a minimum period of three years

- Providing inputs for TIPA activities

- Conducting demonstration and adaptive research on topics related to the project

- Conducting applied and adaptive research to identify best vegetables varieties

A Service Center is vital for the success and the sustainability of **large-scale** development programs, particularly those that deal with innovation where capacity building and organization are a necessity.

3. Organization

Organization is another essential element for the success of a development program. One should always try to work with existing farmers organizations that will on one hand participate in "bringing the message" to the farmers, and on the other hand, organize the farmers to implement the introduced crops and technologies in the best manner. Local farmers' organizations also deal with aspects of marketing and pricing of the products. For example, the onion growers association of Niger many times fixes the minimum price of onions, thus preventing the exploitation of farmers by traders.

New approaches will require the establishment of new organizations. For example, in a project that I advised, we trained selected farmers to produce quality seeds of field crops and vegetables. These local seed producers were then encouraged to create local organizations of seed producers. The local organizations joined hands with other local organizations to create regional organizations. The seed producers' organization will, as in the case of onions, set up prices for seeds. The benefits of this control will help prevent exploitation by seed brokers. They will conduct negotiations with the private sector on seed prices and help organization members in obtaining loans.

Whenever a group of producers is sharing water, land and other means of production, it is essential that they organize themselves in order to enable the smooth operation of the shared resources. For example, women practicing BDL communal AMG or any other activity are regularly organized in an association led by an elected president, a secretary and a treasurer. Various committees such as technical or credit committees are arranged. The whole association meets frequently to discuss problems and decide on activities. Such organization facilitates the collection of dues from each member of the association to be used for investment, materials purchase and extension of micro-credit to group members.

4. Human competency

Human competency is an element that "makes or breaks" a program. Therefore, staff hired by a program should be highly professional, motivated and dedicated. In fact, the problem with most agricultural development programs is the lack of professionalism and staff motivation. Lack of motivation leads to poor performance, and many times, it can result in lack of transparent behavior. Management experts agree that the best way to encourage motivation is to give the staff a sense of ownership in a project and encourage them to offer their own solutions to the problems that they encounter in the field.

> We started working with a farmers' cooperative in the region of Gaya, Niger to install a five-hectare African Market Garden. The equipment and its installation were provided by the project. Farmers were asked to prepare the land and dig the trenches for the distribution pipes as their contribution to the project. But they refused to do it, asking to be paid by the project for their labor.
>
> In a meeting with members of the cooperative, I was told that, so far, all previous projects in their region had paid for labor, so the expectation was that the farmers should be paid this time as well. After giving a patriotic speech, including a threat to move the project to another site, farmers reluctantly agreed to work for themselves without pay.
>
> On the other hand, in the region of Kalale, Benin, where female-managed market gardens were installed, women willingly cut down huge trees, leveled the land and dug the trenches without being paid for their labor. And in the end, they were full of thanks for the project that has provided them with so much income. This group of women was never **spoiled** by projects that paid for labor.

Village animators are a component of many development programs. In each village, an "animator" who is the link between the project and the farmers is appointed. Many of these animators are real idealists who want to see their corresponding villages benefit and are there in the village fields every day of the week working with farmers. While a project lasts, these animators receive some compensation from the project. They also receive professional training including the provision of production manuals. The project then furnishes the animators with a range of means to increase their income. For example, they are taught how to install and maintain irrigation systems; in irrigation projects, they are given irrigated plots, etc. And the project ensures that farmer groups will continue to benefit from the technical guidance of the animator once the official project ends.

These animators actually replace government extension services in places where these services are not effective. The animators are responsible for scaling up the crops and technologies introduced by the project after the project formally ends. They will do this job best if they have an economic interest in the up-scaling of the technologies.

5. Financing development

Financing development in Africa is a very complex and controversial issue. At the micro level many NGO- and government-financed programs tend to pay farmers to participate in the introduction of new technologies. They not only pay the set-up costs but also the labor of the farmers involved in the project. And as happens far too often, the project fails after the "project ends." Farmers then wait patiently for the next project to come along in order to be able to "milk" the new donors.

This attitude has created a sort of parasitism among farmers, as is exemplified in the two anecdotes given in the accompanying text boxes.

Based on this experience, some donors and projects have gone to the opposite extreme. They are now asking the farmers to pay in full for the technologies that are to be introduced. One way to do this is by connecting farmers to micro-credit organizations that will loan them the resources needed for development. Unfortunately, many times when farmers agree to accept these loans, they are not able to pay them back on time because of the short credit terms. Because these micro-credit organizations are taking risks that established banks would not dare take, they compensate by requesting interest rates sometimes reaching as high as 30%. Apparently micro-credit enterprises are becoming very good business because micro-credit institutions are popping up throughout Africa like mushrooms after the rain.

A World Vision friend working in Gaya, Mali once told me the following story:

"I decided to conduct an alphabetization program for women in my region. I brought in teachers and teaching materials and waited for the women to come. But none came. When I asked the women why, their answer was straightforward: "You pay us we will come."

The assumption that projects fail if farmers do not heavily invest in them is not always correct. Projects also fail due to *other reasons* such as the lack of sufficient expertise by farmers, lack of organization, poor value chain, lack of maintenance, short project life, lack of marketing strategy, poor markets, cultural constraints and lack of profitability among many other reasons. *If a farmer realizes that a technology is bringing him a good income, he will continue to implement it and expand it* **even if the specific technology was originally fully financed by a project** (see the box of the "transformation of the Sadore village" below).

When deciding what component of any project should be financed by the project and what component should be financed by farmers, *reason should prevail.* If project managers want to introduce new technologies or new crop species or varieties, they should not expect farmers to take the risk for something that they have not tried before. In this case, the project *should subsidize these new technologies and crops.* Only if and when these technologies succeed and farmers are ready to adopt them should they be requested to finance the up-scaling.

Many African governments that want to encourage industries such as tourism or telecommunication give these industries generous subsidies. But to the best of my knowledge, in many countries, there is no such policy regarding agricultural development. Since this is the main industry of most African countries, it certainly makes sense that governments should subsidize agriculture and

210 | Dov Pasternak

agriculture-based industries as much or more than they do the Hiltons or the Orange cellular company.

To support agricultural development governments should invest not only in roads but in other infrastructure items such as water supply, electricity for water pumping, irrigation systems and telecommunications as well as in subsidized loans, market outlets for products, market information services, extension, agricultural education, and more.

*Agriculture for **small** African producers will not move forward without the readiness and the ability of governments to streamline agricultural development including injecting massive financial support while emphasizing the creation and expansion of the private sector.*

Planning development

Lack of agricultural development in African countries has created a vacuum that has been filled by international and national NGOs, by UN agencies and by many bilateral aid agencies. Each one of these organizations has its own development agenda with very little coordination with governments and with practically no coordination with each other. The result of this situation is duplication of efforts that sometimes are antagonistic to each other. For example, it is possible to find a number of development organizations conducting vegetable production activities in the same village at any given time. Some organizations provide seeds to the farmers, whereas others teach farmers how to produce them. Some organizations do not give training to farmers, while others do. Likewise, in the area of land reclamation, some organizations ask farmers to build water harvesting structures while others pay farmers to do this.

Governments have little control on the direction and shape of agricultural development of their countries and duplication of efforts abound. There are notable exceptions such as the government of Ethiopia which is bringing together major projects from various organizations under one umbrella while retaining the decision making role, but most of the entities involved are taking barely any steps to change this situation.

I have no definite opinion about the need for central planning of agricultural development. Perhaps this is a result of my experience in Argentina, where in each province, there is a planning department employing a significant number of people that produce many documents *that are rarely implemented*. Even though Israel is very proud of its comprehensive rural planning approach, I know of a large fertile region in the western Negev Desert where the settlement authorities planned their agricultural activities on grapes for wine production. This approach failed, resulting in mass desertion of newly constructed villages by their inhabitants. Only when a group of entrepreneurs without any "planning" had success demonstrating that this region was very suitable for the production of out-of-season vegetables did the area prosper.

Central planning can reduce costs and improve performance. On the other hand, if it is based on the wrong assumptions on wrong doctrines and it does not consider farmers wishes, it can lead to failure. Still, **each particular** project requires planning; otherwise, it cannot proceed and succeed. However, one must anticipate that slight or even considerable deviations from the original planning will take place during the execution of the program either because the original plan proves to be faulty or because better solutions than those planned are discovered while the project is progressing. Planners should, therefore, be extremely flexible, rejecting approaches found to be unsuitable *and be willing to accept new approaches not originally envisioned.* This suggestion is problematic for bureaucrats to incorporate as it is their job to evaluate the success of any particular program based on indicators embedded in the original plan. To solve this dilemma, I suggest those responsible for project evaluation create very **general** "Indicators of Success," concentrating more on general objectives and less on specific ones.

Additionally, central planning by governments can be harmful to the farmer because political considerations are often involved. As I have already mentioned, providing food to the urban population at low prices is a major priority of governments because increases in food prices invariably results in "food riots" in cities. To prevent this, governments plan massive production of cheap food by farmers, which is in direct contradiction to farmers' interests and, in the long term, also to national interests.

The final question in this chapter is how to implement development projects. Before delving deeply into this question, I must issue a warning: A development project must be managed by one single entity. This entity controls the project by controlling the budget. It is possible to employ as many partners as needed, but all partners must depend on one manager for the project.

Examples:

- The Desert Margins Initiative of UNEP was coordinated by ICRISAT and executed by many African NARs and other CGIAR Centers. Each partner coordinator received "his share of the pie" and ran his country plan quite independently from the coordinating office. The lack of control of these partners was a disaster. Resources were not used for the efficient execution of project plans and, accordingly, results were very poor.

- I was once involved in a project in Morocco financed by USAID, aimed at cooperation in agriculture between Israel, the US and several Arab countries. The project was administered by the Moroccan partner, a shrewd businessman who used this as an opportunity to beef up irrigation and farm machinery on his private farms at the expense of the American tax payer.

In contrast to previous examples, a USAID project for promotion of the African Market Garden in Burkina Faso under my responsibility was carried out in cooperation with the local NAR called INERA. INERA hired the technicians, but the project was run from ICRISAT-Niger, who transferred salaries on a quarterly basis to INERA. All other expenses were paid for directly by ICRISAT. The two technicians were fully under the technical supervision of the ICRISAT coordinator.

This strategy had an added benefit in fostering excellent relationship building between ICRISAT and INERA at that time.

The major points to consider in planning large development projects are:

- Sustainability and the likelihood of up-scaling without external help

- Massive and effective transfer of know-how to the producers

- Technical support of the project through Service Centers

- Maximum administrative control of project funds by the central entity that runs the project

- Development of the private sector

- Effective involvement of research

Once planning is done, the next question concerns how to execute a development project. There are actually many ways to do this. Here, I present a successful model adopted by the NGO called CLUSA with some modifications of my own:

At the top of the pyramid, there is a "coordinator," who in practice is the overall director of the project. Underneath him, there is a team of administrative personnel (administrator, purchasing officer, accountant, consultants, engineers, monitoring and evaluation team, drivers, etc.) and regional coordinators. Under the regional coordinators, there are "facilitators" who receive salaries from the project, are housed in the project villages and have access to all project recipients, particularly to village and regional leaders.

Next are the "project technicians" (also under the coordinators) who are responsible for training and technology transfer. And last but not least are the village animators who supervise the daily work in the project fields. The animators are the contact point between the coordinator, the facilitators, the technicians and the farmers.

If the project deals with technology transfer, the project technicians are the most essential component of the team while the project is running since they are in charge of the *transfer of know-how.* The facilitators are in charge of organization and rank second in importance after the technicians.

Any project should try to partner with professional organizations with capacities needed by the project that are not available at the executing agency. The most important partners however are regional farmers' associations that facilitate technology transfer, guarantee sustainability and promote successful technologies. And last but not least is the involvement of the private sector including contractors of all kinds, input suppliers, credit suppliers and traders, among others.

The Sadore village is situated opposite the ICRISAT Research Station in Niger. Until 2003 this was a typical Nigerien village. People lived in traditional igloo shaped mud houses. Household livelihoods depended on rainfed annual crops and on occasional labor at the ICRISAT station. Women carried out their traditional chores of firewood collection, food preparation and looking after their children. They were totally depended on their husbands for their livelihoods. Starting 2003 ICRISAT introduced to the Sadore women a series of activities such as market gardening, the BDL and a fruit trees nursery. The women were followed up for 7 years and then were left on their own.

And an amazing thing happened. Every village household established a fruit trees nursery at the back of their house. The whole village is now producing fruit trees plants, men and women. Women are earning now $2,000 per year as compared with the country average of $500/year. About 50 new spacious houses were constructed in the village. Most children go now to secondary school in a nearby town. And after school they go to the University or college. Women are now buying land and give these to their relatives to work on. They are buying livestock and more. Marital relations have improved dramatically. Sadore women have moved their village out of poverty. There is no way back. This experience must be repeated!

Oumou is also part of a women's fruit tree nursery. Each of the 30 members of this nursery earns about USD $2,000 a year, five times the average income in Niger. According to Oumou:

"With the BDL and nursery activities, I have a good income. This means I can clothe and educate my children. I have my own mobile phone and have also bought a few sheep. Last year was a drought year, but because of the income that I brought home, my family did not suffer from hunger. Many women in my association have built new mud brick houses to replace the small and crowded traditional village houses. I can definitely say that the status of all the women in our association has changed. We are less dependent on our husbands and *we are more respected by them* as we contribute to family expenses."

Most partners, together with project members, meet periodically to receive updates on the project and to plan activities on a quarterly or bi-annual basis. During the project period and after project financing ends, the village animators, the farmers' organizations and the private sector assume responsibility of up-scaling of the project.

Womens' roles in agricultural development

In Africa, women play a major role in daily agricultural activities. They help their husbands plant their fields, weed and harvest the crops and look after the small household animals, in addition to caring, in many instances for small irrigated vegetable plots during the off-season. However, these women are often marginalized by their societies. They do not have tenure over the land nor are they allowed ownership of agricultural utensils. They are discriminated against in education and suffer from higher illiteracy rate than men.

Even so, it is widely accepted that *gender equality* in African villages, particularly in dry West and Central Africa could result in boosting agricultural development (ADFVI Progress Report 2008).

Many emphasize involvement of women in politics as an important step towards womens' equality. This is indeed an effective way to empower women, but it is not as effective as providing women with income. When women earn as much as or more than men, it becomes very difficult to continue to discriminate against them at the family and community levels.

Therefore, one of the major duties of development, particularly agricultural development, is to significantly increase womens' income. Another way to empower rural women is to give them ownership over land. Projects such as the BDL and the AMG are classic examples of measures that significantly empower women through land ownership.

The boxed story above describes the amazing transformation of the Sadore village that was carried out through women empowerment.

Conclusions

Rabbi Hillel was a prominent Jewish scholar at the time of the Roman occupation of the kingdom of ancient Judea. One day, this great teacher was approached by a gentile who asked if the rabbi could teach him the whole Jewish Torah while he was standing on one leg.

"No problem," answered Rabbi Hillel. "Please stand up on one leg and repeat after me: That which is hateful to you, do not do to your fellow" which the gentile did.

"Now," said Rabbi Hillel, "that you know the essence of the Torah, all the rest is commentary. You can get back onto your two legs and go to learn the rest."

My "Rabbi Hillel message" in this book is: Please treat agriculture first and foremost as a *business* where know-how, relative advantage, innovation, marketing, financing, organization, high value crops, irrigation, post-harvest treatment etc. are used to maximize profits under each specific set of conditions. Then, and only then, look at agriculture as a pipeline for transferring your visions of food security, organic agriculture, child nutrition, safe water, combating climate change and other development actions.

If your vision goes hand in hand with the promotion of the agricultural *business,* then you are in good shape. But if it clashes with the business aspects, then please **keep your vision out of it.**

Now you can stand on your two legs and start developing agriculture in dry Africa as it should be done.

Bibliography

- Agmon, R. 1999. *Tropical Africa agricultural and rural development* (in Hebrew). Hansen Printing.

- Alston, J. M., Steven, D., and Philip, G.P. 2006. "International initiatives in agriculture R&D: The changing fortunes of the CGIAR." In, Pardey, P. G., Alston, J. M., and Piggott, R. R., (eds.). *Agricultural R&D in the Developing World: Too Little Too Late?* Washington, DC: IFPRI, pp.313-360.

- Anderson, K. 2006. Reducing distortions to agricultural incentives: Progress pitfalls, prospects. *American Journal of Agricultural Economics* 88: 1135-1146.

- Bauer, P. 1957. *Economic Analysis and Policy in Underdeveloped Countries.* Duke University Press: Durham, N.C.

- Baidu-Forson, J. 1999. Factors influencing adoption of land-enhancing technology in the Sahel: Lessons from a case study in Niger. *Agricultural Economics* 20: 231-239.

- Belder, P., Rohrbach, D., Twomlow, S., and Senzanje, A. 2007. Can drip irrigation improve the livelihoods of smallholders? Lessons learned from Zimbabwe. Global Theme on *Agroecosystems Report* No. 33. International Crops

Research Institute for the Semi-Arid Tropics: Bulawayo, Zimbabwe.

• Borochov-Neori, H., Judeinstein, S., Greenberg, A., Fuhrman, B., Attias, J., Volkova, M., Hayek,T., and Aviram, M. 2008. Phenolic antioxidants and antiatherogenic effects of marula (*Sclerocarya birrea* subsp. *caffra*) fruit juice in healthy humans. *Journal of. Agricultural and Food Chemistry.* 56: 9884–9891.

• Breman, H. and Kessler, J. J. 1999. The potential benefits of agroforestry in the Sahel and other semi-arid regions. *Development in Crops Science* 25: 39–47.

• Buerkert, A., Michels, K., Lamers, J. P. A., Marshner, H., and Bationo, A. 1996. Anti-erosive, physical, soil and nutritional effects of crop residues. In, B. Buerkert, B.E. Allison and M. von Oppen (eds.). *Wind Erosion in Niger.* Kluwer Academic Publishers in cooperation with the University of Hohenheim. Dordrecht, The Netherlands. pp. 123–138.

• Buerkert, A., Lamers, J. P. A., Marschner, H. and Bationo, A. 1996a. Inputs of mineral nutrients and crop residue mulch reduce wind erosion effects on millet in the Sahel. In, B. Buerkert, B.E. Allison and M. von Oppen (eds.). *Wind Erosion in West Africa: The problem and its control.* Proc. Intern. Symp., Univ. Hohenheim, Stuttgart, Germany, 5-7 Dec. 1994. pp. 145-160.

• Burdon, D. J. 1985.Groundwater against drought in Africa, In, *Hydrogeology in the Service of Man: Mémoires of the 18th Congress of the International Association of Hydrogeologists,* Cambridge, IAHS Publication No. 15.

• J. Burney, L. Woltering, M. Burke, R. Naylor, and D. Pasternak 2010. Solar-powered drip irrigation enhances food security

in the Sudano–Sahel. www.pnas.org/cgi/doi/10.1073/
pnas.0909678107. 6 pages

- DGIRH 2001. *Etat deslieux des resources en eau du Burkina
  fasoet de leur cadre de gestion.* Ouagadougou, Burkina Faso:
  Direction Génerale de l'Hydraulique, Minister de l'Agriculture,
  des l' Hydraulique et des Ressources Halieutiques.

- Diatta, M., Diack, M., Sene, M., and Pasternak, D. 2001.
  Bioreclamation of saline soils of the west coast of Senegal
  2001. In, D. Pasternak and A. Schlissel (eds.). *Combating
  Desertification with Plants.* Kluwer Academic/Plenum
  Publishers, New York, pp. 329-38.

- Drenches, P., Graeme, S., Soon, M., and Coffee, O., 2006.
  Informal Irrigation in Urban West Africa: An Overview. IWMI
  Research Report 102, International Water Management
  Institute: Colombo, Sri Lanka. http://www.iwmi.cgiar.org/
  pubs/ pub102/RR102.pdf

- Dagger, C.W. 2007. Ending famine, simply by ignoring the
  experts. *New York Times,* December 2, 2007.

- Easterly, W. 2006. *The White Man's Burden: Why the West's
  Efforts to Aid the Rest Have Done So Much Ill and So Little
  Good.* The Penguin Press: New York.

- Fan, S., Johnson, M., Sourer, A. and Mugabe, T. 2008. Investing
  in Africa to halve poverty in 2015. *IFPRI Discussion Paper
  No. 0075.*

- FAO 2009. How to feed the world 2050: The special challenge
  for sub-Saharan Africa. *FAO High Level Experts Meeting.*
  Rome, 12-13/10/09.

- Fatondji, D., Pasternak, D., and Woltering, L. 2008. Watermelon production on stored rainwater in Sahelian sandy soils. *African Journal. Of Plant Science.* 2: 151-160.

- Fuglie, L. J. 1999. The miracle tree: Moringa oleifera, natural nutrition for the tropics Published by Church World Service

- Gates Foundation 2011. *Agricultural Development in Africa: Fact Sheet.* The Bill and Melinda Gates Foundation Global Development Program.

- Hoagland, E., Ndjeunga, J., Snook, L., and Pasternak, D. 2011. **Dryland tree management for improved household livelihoods: Farmer Managed Natural Regeneration in Niger**. *Journal of Environment Management* 92: 1696-1705.

- Harrison, P. 1987. *The Greening of Africa. Breaking Through in the Battle for Land and Food.* Paladin Grafton Books: London.

- Hassan, A. 2006. Improved traditional planting pits in the Tahoua department, Niger. In, Rein, C., Scions, I. and C. Tooling (eds.) *Sustaining the Soil: Indigenous Soil and Water Conservation in Africa.* Earth scan, London, 56-61.

- Hussein, I. and Hanja, M.A. 2004. Irrigation and poverty alleviation: Review of the empirical evidence. *Irrigation and Drainage* 53: 1-15.

- ICRAF, 2009. *Agroforestry: A Global Land Use.* World Agroforestry Center, Nairobi, Kenya.

- Marenya, P. P. and Barrett, C. B. 2009. State conditional fertilizer yield response on Western Kenyan soils. *Am. J. Agric. Econ.* 91: 991-1006.

- Nikiema, A., Pasternak, D., Fatondji, D., Senbeto, D., Ndjeunga, J., Woltering, L., and Abdoussalam, S. 2008. Fruit trees for the Sudano-Sahel region of West Africa. *Chronica Horticulturae* 48: 24-29 D.

- Paarlberg, R.L. 2009. *Starved for Science: How Biotechnology is Being Kept Out of Africa.* Harvard University Press.

- Pasternak, D. 2013. Lablab (*Lablab purpureus*) a new staple crop for the Sudano Sahel. Echo Technical Note #73

- Pasternak, D. and Bustan, A. 2003. The African Market Garden. In Stewart, B.A. and Howell, T. (eds.). *Encyclopedia of Water Science.* . Marcel Dekker Inc.: NY, pp.9-15.

- Pasternak D., Nikiema A., Fatondji D., Ndjeunga, J., Koala, S. Dan Gomma A., and Abase T. 2005. The Sahelian Eco-Farm. In, Omanya, G. and Pasternak, D. (eds.) *Sustainable Agriculture Systems for the Drylands.* ICRISAT: Patancheru, 286-296.

- Pasternak, D., Nikiema, A., Senbeto, D., Dougbedji, F., and Woltering L. 2006. Intensification and improvement of market gardening in the Sudano Sahel

- region of West Africa. *Chronica Horticulturae* 46: 24-28.

- Pasternak, D. Senbeto, D., Nikiema, A., Kumar, S., Dougbedji, F., Woltering, L., Ratnadass, A. and Ndjeunga. J. 2009. Bioreclamation of degraded lands with women empowerment in Africa. *Chronica Horticulturae* 49: 24-27.

- Pasternak, D., Housseini, I. and Drori, U. 2013. The Arziki Onion Store: New, Effective and Affordable Onion Storage for Small Producers. *Chronica Horticulturae* 53: 12-15

- Polak, P., Nanes, R., and Adhikari, D. 1997. A low cost drip irrigation system for small farmers in developing countries. *Journal of the American Water Resources Association* 33: 119-124.

- Punjab National Bank 2010. *Wheat Basic.* Special report by the Punjab National Bank.

- Sachs, J. D., 2005. *The End of Poverty: Economic Possibilities for Our Time.* The Penguin Press. New York, NY.

- Valenza, A., Grillot, J. C. and Dazy, J. 2000. Influence of groundwater on the degradation of irrigated soils in a semi-arid region, the inner delta of the Niger River. Hydrology Journal 8: 417-421

- Woltering, L., Pasternak, D., and Ndjeunga, J., 2011a. The African Market Garden: The development of an integrated horticultural production system for smallholder producers in West Africa. *Irrigation and Drainage* (Wiley) DOI: 10.1002/ird.610.

- Woltering, L., Ibrahim, A., Pasternak, D., and Ndjeunga, J. 2011b. The economics of low pressure drip irrigation and hand watering on vegetable production in the Sahel. *Agricultural Water Management* 99 pp. 67-73.

- Weitz, R. and Rockach, A. 1968. *Agriculture Development: Planning and Implementation (Israel case study).* R. Reidel Publishing Company: Dordrecht-Holland.

- Wilson, E.O. *Letters to a Young Scientist* , TEDMED Talk. http://www.ted. com/talks/e_o_wilson_advice_to_young_scientists.html, retrieved 3/8/12.

- World Bank 2008. Development of irrigation in Niger: Diagnostic and strategical options (in French). Agriculture and rural development AFTAR. African region

- Young, A. 2007. *Agroforestry for Soil Management.* ISBN 0-85199-189-0

Appendixes

Appendix 1

List of vegetable varieties for the three growing seasons of the Sudano Sahel

Varieties marked with an asterisk were selected and/or developed by ICRISAT-Niger. VDG means Violet de Galmi. Foundation seeds of most varieties can be obtained at masalifou@yahoo.com.

Dry-Cool		Dry-Hot		Rainy	
Species	Variety	Species	Variety	Species	Variety
Onions	VDG*¹	Melons	Ein Dor*	Sweet corn	True gold*
Potatoes	Desire	Okra	Konni*	Okra	Konni*
Carrots	Uberl.*	Eggplant	Black beauty	Tomatoes	Icrixina*
Cabbage	Oxylus	Cucumbers	Bet Alpha*	Cucumbers	Bet Alpha*
Lettuce	Maya*	Moringa	PKM 1*	Lettuce	Maya*
Sweet peppers	Diffa*			Eggplant	Black beauty
Hot peppers	Safi*			Moringa	PKM 1*
Tomatoes	Sadore*				
Moringa	PKM 1*				

Appendix 2

List of successful fruit tree varieties at the Sadore Research Station of ICRISAT

For further information and the acquisition of plants contact: s.abdoussalam@icrisatne.ne or masalifou@yahoo. com.

Family	Genus	Species	Common name	No. of varieties / provenances
Anacardiaceae	Mangifera	*indica*	Mango	27
Anacardiaceae	*Sclerocarya*	*birrea sub."caffra"*	Marula	4
Anacardiaceae	*Lannea*	*microcarpa*	Wild grape	1
Apocynaceae	Saba	senegalensis		7
Arecaceae	Phoenix	*dactilifera*	Dates	10
Caricaceae	*Carica*	*papaya*	Papaya	1
Cesalpiniaceae	*Tamarindus*	*indica*	Sweet Tamarind	4
Moraceae	*Ficus*	*carica*	Fig	5
Moraceae	Morus	*alba*	White Mulberry	1
Moringaceae	*Moringa*	*oleifera*	Moringa	4
Punicaceae	Punica	*granatum*	Pomegranate	9
Rhamnaceae	*Ziziphus*	*mauritiana*	Pomme du Sahel	10
Rhamnaceae	Ziziphus	*spina christi*	Christ thorn	
Rhamnaceae	*Ziziphus*	*rotundifolia*	-	
Rutaceae	*Citrus*	*grandis*	Pommelo	3
Rutaceae	*Citrus*	*paradisixgrandis*	Sweety	1
Rutaceae	Citrus	*tangelo*	Tangelo	4

Family	Genus	Species	Common name	No. of varieties / provenances
Rutaceae	Citrus	reticulata	Mandarin	1
Rutaceae	*Citrus*	*sinensis*	**Sweet Orange**	4
Rutaceae	*Citrus*	*aurantiifolia*	**Lime**	3
Rutaceae	Citrus	*limon*	**Lemon**	1
Rutaceae	*Citrus*	*aurantium*	**Bigaradier**	1
Rutaceae	Citrus	volkameriana		1
Rutaceae	*Citrus*	*hybrid*	**Idit**	1
Sapotaceae	*Manilkara*	*zapota*	**Sapodilla**	4
Vitaceae	*Vitis*	*vinifera*	**Grape**	9
Total varieties				114

www.ingramcontent.com/pod-product-compliance
Lightning Source LLC
Chambersburg PA
CBHW042146220326
41599CB00003BB/5